"Big data, machine intelligence, digital age ar
daytime meeting or evening conversation. In t
and succinctly challenges the organizations of
efficient. And lays out a roadmap for them to succeed. A must read for CEOs, CXOs, consultants and academics who embrace change and are true leaders."

Dr. Sanjiv Chopra MD, MACP. Professor of Medicine, Harvard Medical School

"Dr. Srinivasan talks to us from a future that he has already seen and in many ways realized through practical applications he describes in his book. This is a breakthrough treatise on Artificial Intelligence in the virtual or non-physical world of Business Processes, de-mystifying, deconstructing and making the cold logic and magic of AI accessible to all. This is a must read for anyone curious about how work will get done in the future, so you can start making informed choices today."

Joy Dasgupta, SVP, RAGE Frameworks

"For a business leader to constantly deliver superior business performance is daily fodder. The challenge lies in driving change in organizational behaviour. Dr. Srinivasan shifts the paradigm; he provides solutions for what may hitherto have been impossible or prohibitive!"

Sanjay Gupta is CEO, EnglishHelper, Inc. and formerly SVP, American Express

"This book is a must read for business and technology leaders focused on driving deep transformation of their businesses. Venkat has brilliantly outlined practical applications of intelligent machines across the enterprise. The best part, this eloquent narration is based on problems he has solved himself at RAGE Frameworks."

Vikram Mahidhar, SVP, RAGE Frameworks

"An amazing book addressing the challenges faced by all businesses. Having gone through these challenges myself in my professional career with several global organizations I can totally relate to the book. Business needs are changing at a very fast pace and Dr. Srinivasan has offered very practical solutions. Process oriented solutions are flexible and allows business to adapt quickly to these fast changing requirements. Intelligent automation has the ability to dramatically transform organizations and provide a competitive edge. A must read for business leaders."

Vivek Sharma, CEO, Piramal Pharma Solutions

"Technology is intended to make business more agile, more efficient. But time and again, this same technology becomes a straitjacket once implemented, and forces the business to adapt, instead of the other way around. The book provides a step-by-step deconstruction of what it takes to be agile, efficient and intelligent. Based on this deconstruction, Venkat develops an alternate architecture that leads to the truly agile, efficient and intelligent enterprise. This is not just theory and concept, but implemented and running at several leading global corporations today. Ignore at your own peril!"

Deepak Verma, Managing Director nv vogt and formerly, CEO, eCredit, Inc.

"Over the last 30 years I've helped a number of companies grow faster than their competitors in many industries. But, in so many ways, the enterprise of today has changed: it's global, its customers have many new expectations for service, it is facing new competition from new business models, and it has a new workforce with different skills and desires. Wherever you sit in this new corporation, Srinivasan gives us a practical and provocative guide for rethinking our business process...using data and user controlled access as a speedy weapon rather than a cumbersome control and calling us all to action around rapid redevelopment of our old, hierarchical structures into flexible customer centric competitive force. A must read for today's business leader."

Mark Nunnelly, Executive Director, MassIT, Commonwealth of Massachusetts and Managing Director, Bain Capital

"'Efficiency', 'agile,' and 'analytics' used to be the rage. Venkat Srinivasan explains in this provocative book why organizations can no longer afford to stop there. They need to move beyond – to be 'intelligent.' It isn't just theory. He's done it."

Bharat Anand, Henry R. Byers Professor of Business Administration, Harvard Business School

"Venkat Srinivasan is one of those rare individuals who combines the intellectual horsepower of an academic, the foresight of a visionary, and the creativity of an entrepreneur. In this book he offers a compelling vision of the next generation of organization—the intelligent enterprise—which will leverage not just big data but also unstructured text and artificial intelligence to optimize internal processes in real time. Say good-bye to software systems that don't talk to one another and cost a fortune to customize, and say hello to the solution that may become the new normal. If the intelligent enterprise seems utopian, read the chapters on how some companies have actually applied this concept with impressive results. Let Srinivasan give you a peep into the future. This is a must-read book for CEOs and CTOs in all industries."

Ravi Ramamurti, D"Amore-McKim Distinguished Professor of International Business & Strategy, and Director, Center for Emerging Markets, Northeastern U.

"Dr. Venkat Srinivasan has written a book aimed at business professionals and technologists. This is not geek speak, not an academic treatise. Venkat writes with great clarity and precision based on his real-life experience of delivering solutions through the RAGE AI platform. It is about the brave new world that narrows the gap between technology and business. Most of us have labored with technology projects that took too long, cost too much and delivered less than expected. Process-oriented software and Artificial Intelligence can create solutions that are flexible, smart and efficient. The book has practical advice from a thoughtful practitioner. Intelligent automation will be a competitive strength in the future. Will your company be ready?"

Victor J. Menezes, Retired Senior Vice Chairman, Citigroup

THE INTELLIGENT ENTERPRISE IN THE ERA OF BIG DATA

VENKAT SRINIVASAN

WILEY

Published by John Wiley & Sons, Inc., Hoboken, New Jersey.
Published simultaneously in Canada.

For general information on our other products and services or for technical support, please contact our Customer Care Department within the United States at (800) 762-2974, outside the United States at (317) 572-3993 or fax (317) 572-4002.

Wiley also publishes its books in a variety of electronic formats. Some content that appears in print may not be available in electronic formats. For more information about Wiley products, visit our web site at www.wiley.com.

Library of Congress Cataloging-in-Publication Data applied for.

ISBN: 9781118834626

Printed in the United States of America

10 9 8 7 6 5 4 3 2 1

This book is dedicated to my family

To my wife Pratima from whom I have learned so much, for her unwavering support for my efforts, and to our daughters, Meghana and Nandini, with love and gratitude

To my parents, Srinivasan Varadarajan and Sundara Srinivasan, for braving significant challenges in their lives and insulating me from them so I could pursue my dreams

CONTENTS

PART II AN ARCHITECTURE FOR THE INTELLIGENT ENTERPRISE

PREFACE

Two centuries ago, Adam Smith laid down architectural principles that govern how enterprises organize themselves and function. In this book, we present an entirely new way to think about how enterprises should organize and function in the digital age. We present the enterprise architecture for the future: an architecture that recognizes the power of the emerging technology environment, enables enterprises to respond rapidly to market needs and innovation, and anticipates such needs by sensing important developments in internal and external environments in real time.

Enterprises continually strive toward becoming efficient and competitive through various means. Prompted by the TQM and radical re-engineering movements of the 1980s and 1990s, many enterprises have attempted to embrace process orientation as the key to efficiency and competitive differentiation. However, most have had only limited success in becoming process efficient. This may be largely because in today's dynamic business environment, the static and unresponsive nature of most technology paradigms has stifled any significant progress. In recent years the flood of digital information, called *big data*, has compounded this challenge and opened yet another front for businesses to factor into their strategies.

Most enterprises are severely constrained by their inability to change their processes in response to market needs. Despite all the attention toward business process management and process orientation, businesses still struggle with time to market and flexibility issues with technology. Technology instead of enabling such changes has become a serious inhibitor. Changing business processes embedded in software applications is often a lengthy, arduous process replete with cost overruns, missed timelines, and failures. The rapid pace of technology obsolescence has continued to require specialized training and skills and has exacerbated this issue further.

To keep up with business demands, businesses have gravitated toward packaged applications at least for what they perceived to be non-core functions like resource planning and financial accounting. For most enterprises, it is too expensive and difficult to maintain a custom technology application environment. Initially it was widely believed that the new world business order implied standardization of business processes even beyond non-core functions. It was argued that firms would seek to standardize business processes for several reasons – to facilitate communications, enable smooth handoffs across process boundaries, and allow comparative analyses across similar processes. This was hypothesized to revolutionize how businesses organized themselves. But such thinking has resulted in enterprises being forced to operate within the limits of the prevalent technology paradigms.

The Internet phenomenon was still nascent in the late 1980s/early 1990s. Since the mid-1990s, the Internet has become pervasive in businesses and peoples' personal lives; the rate of new information flow has been and is staggering. The rapidly emerging Internet of Things promises to add a whole new dimension of information at an extraordinary scale. If we add the viral spread of social media to an overabundance of information, corporations face an enormous challenge and opportunity to intelligently harness the wealth of knowledge and insight contained in such information.

Yet, over the last decade, the gap between "technology speak" and "business speak" has narrowed considerably. The ability to create and maintain a technology application has got considerably easier. The age of highly flexible process-oriented software frameworks that enable a corporation to configure its business processes, is now available to enterprises. Simultaneously, a whole new class of technologies has emerged to help enterprises deal with the explosive growth in data, and developments in cognitive computing promise a range of capabilities that will enable machines to do much more than be keepers and facilitators of data.

The enterprise of tomorrow has the opportunity to be intelligent in addition to being efficient. It requires the ability to monitor and analyze internal and external threats and opportunities continuously, and to make adjustments in operational processes to counter such threats or leverage opportunities. In doing so, it is not sufficient to analyze the enormous amount of unstructured information that has become available. An intelligent enterprise will need to seamlessly integrate such analytical processes into its normal operational processes. These two worlds are not distinct and dichotomous; rather, they are part of the same continuum. Without integrating these two sets of processes, enterprises will not achieve the desired results. Remember, enterprises are far from having solved the challenge of rapidly adapting their operational processes to the dynamic business environment. Most firms are still struggling to get their myriad systems to talk to each other, data quality issues are still bogging them down, and the list goes on.

These developments portend an enormous change in how enterprises architect themselves and operate. The historical constraints of unresponsive technology paradigms will now be history. By being able to configure technology to suit their business process needs, enterprises will be able to move away from tightly packaged applications without the overhead of custom software maintenance. Coupled with the ability to potentially understand unstructured data in addition to structured data, enterprises have the opportunity to think entirely differently.

Another fact is that today's enterprise architecture is largely people-centric. People have been largely the business process execution glue in an enterprise. In many enterprises people function as the process orchestrators and especially in the knowledge-based industries, people often execute their tasks manually. The time has come for technology to be the process orchestrator in the enterprise, control business process execution, increasingly enabling repetitive tasks to be executed in an automated fashion. Humans will have the opportunity to focus on design and not repeated execution. Flexible software frameworks and the ability to understand the meaning of unstructured documents will provide enormous power to enterprises in designing an entirely new architecture for doing business. This is the central idea of this book.

This book is divided into three parts. Part I frames the challenge enterprises face in greater detail – the challenges of the digital age, the need to adapt to the increasingly dynamic business environment, the inflexibility of systems and the inability to change business processes as needed, the constraints of working within the tight boundaries of packaged applications, the disadvantages of customizing packaged applications thereby rendering their core advantages invalid, and the explosive growth in information and the overload and asymmetry it has created.

Part II outlines an architecture for the intelligent enterprise. How should enterprises architect themselves in the digital age? Has business technology matured enough to allow businesses to configure and re-configure their business processes at will? Are we at a point where businesses can uncommoditize business processes without the overhead of expensive custom software development and maintenance? And how can enterprises systematically harness intelligence from all this data?

First, Chapter 2 delves into efficiency and agility, with focus on the benefits and challenges of a process-oriented enterprise. All of us recognize that labor arbitrage driven outsourcing is clearly not the answer in the long term. The discussion takes you through the current state of business technology and the reasons for why even contemporary software development platforms and methods are not delivering the efficiency and agility enterprises need to be competitive. This may sound surprising, but *agile methodologies will not deliver speed and flexibility* that businesses need. No code model-driven software platforms with an extensive set of model-driven abstract components can address the efficiency and agility challenge. Instead, such a platform can enable near real time, flexible software development and cut typical software development lifecycles to a fraction of what they are otherwise. The chapter discussion walks the reader through a no-code, meta model-driven platform that makes near real-time software development a reality.

Chapter 3 addresses the intelligence dimension with a focus on big data and artificial intelligence. I have intentionally excluded a discussion of computer vision from the scope of this book because of space and time. The chapter presents a taxonomy of AI problems and outcomes to demystify it to the reader. An overview of popular AI solution methods follows. I have tried to balance the treatment between being too technical and yet provide the reader with enough detail to develop a good appreciation for the nature of these methods. By relating these methods back to the taxonomy, I hope the reader will develop an overall understanding of how and where AI is beneficial.

Ninety percent of the content growth on the Internet is unstructured text. Especially as it relates to the handling of natural language, the chapter addresses the important point that most of the current methods, platforms, and tools, including IBM Watson and Google, are based on computational statistics and do not attempt to understand the natural language text at all. The chapter presents the reader with a cognitive intelligence framework that attempts to describe natural language and provide contextually relevant results. Further, there is a trade-off to be made between methods that yield black box solutions and methods that provide traceable, contextually relevant solutions. The cognitive intelligence framework presented in the chapter is not a black box, and its results and reasoning are completely traceable.

Chapter 4 presents an architecture for an intelligence enterprise. The architecture integrates the no-code meta model-driven architectural paradigm for efficiency and agility from Chapter 2 and the traceable cognitive intelligence framework from Chapter 3. The resulting architecture will consist of intelligent machines that learn from humans and data. Fundamentally, I suggest that in the enterprise of tomorrow, the execution aspects of a business will be largely machine run whereby people will be directed by machines and the design aspect of a business will be machine informed as a result of the intelligence gathered by machines. I also review the implications of such an architecture on the current people-centric workplace. Specifically, we revisit the humans versus machines debate and potential impact of the intelligent enterprise on jobs.

Part III presents three real world case studies incorporating the ideas discussed in the previous chapters.

Chapter 5 presents a next-generation architecture for wealth management advisory firms. The wealth management industry is in the throes of a seismic shift with the massive millennial transition, recognition that the historical focus on diversification without explicitly considering investor needs is suboptimal, and the rise of robo-advisors challenging the hegemony of large wire houses. We describe a flexible intelligent framework comprising intelligent machines that can enable wealth advisory firms and advisors to transition to E4.0.

Chapter 6 presents an application to systematically harness real time intelligence to enable active asset managers generate alpha to outperform financial markets. Finding alpha consistently is the Holy Grail in the asset management world. Few sectors in the economy are affected as fundamentally as the investing world with the enormous increase in the availability and flow of information. The application described is a flexible end-to-end solution that includes natural language understanding to process huge amounts of information intelligently and identify possible inefficiencies. Active asset management will move to E4.0 with such an approach.

Chapter 7 explores the use of machine intelligence in the audit profession. This industry is ripe for a major disruption. The fiduciary audit and assurance process is largely manual today and has not changed much since my days as an auditor in the late 1970s. The solution, as presented in the chapter, is an intelligent architecture for the audit firm.

As I show in this book, today there is a fundamentally transformative opportunity to leverage technology like never before in architecting a digital transformation of any enterprise. The opportunity will soon become an imperative. It is my hope that the central ideas of this book will help the business or technology leader see the enormous possibilities for change. The real solutions and options that illustrate this thesis are presented through case studies that demonstrate how to realize these possibilities.

ACKNOWLEDGMENTS

This book is about a big, broad topic and has been in the making for at least two decades. It is the reflection of a lot of learning from colleagues, customers, teachers and friends.

I got the computing bug in the late 1970s working at a large US multinational in Delhi, India. I used to hang around the freezing cold area of the office floor where a couple of IBM 1401s were housed along with all the card punching and reading machines! Later I learned that those machines were already dinosaurs here in the United States, but they were operated with awe back in India those days. I was not trained as a computer programmer but bribed my way into the computer center by helping several programmer friends with punching and running the cards through the readers. From those days to now Internet, tablets, and smart phones, I have witnessed an incredible rate of technology advance in my lifetime to date, and the pace of acceleration seems to be only gaining even more momentum!

Just as Warren Buffett famously talks of his ovarian lottery, I feel incredibly lucky and privileged to have had the ability to learn the way I did and for the breaks and opportunities that came along the way to shape that learning and my professional journey. There are so many that I owe a deep debt of gratitude to. Thanks to my dear friend, Dr. Sanjiv Chopra, I am reminded of Captain Charlie Plumb and his deeply incisive "who packed your parachute" parable as I think back to the times and people who have helped me get to where I have.

I would like to start by thanking my manager at the US multinational who took a chance with me in a significant role as Cash Manager, which got me initiated with my love for management and data science. I had the freedom to solve numerous operational challenges that I believe created in me a self-belief to innovate and solve problems however difficult they might seem.

My advisor at the University of Cincinnati, Professor Yong H. Kim, apart from being an accomplished academic, a patient and wise mentor, had the fortitude and courage to deal with an unconventional doctoral thesis combining finance and expert support systems. I learned a great deal at the University of Cincinnati from some incredibly brilliant teachers who taught me rigorous methods of scientific inquiry and problem solving, apart from teaching me subject matter expertise.

My six years at Northeastern University were very fruitful. I benefited greatly from an environment that was conducive to research and was fortunate to work with a group of like-minded colleagues who were all so passionate about their respective fields of research and so wonderfully collaborative. I would single out the late Professor Jonathan Welch, Finance Department Head at that time, Professor Paul Bolster, and the late Thomas Moore, my Associate Dean, for their encouragement and support.

The roots of my entrepreneurial journey were sown a fateful day in April 1985 when I returned a call from Norm Thomson, then a senior executive at Procter & Gamble. What ensued was a series of research projects that evolved into consulting assignments and eventually, I decided to turn an entrepreneur. I learned a lot from watching Norm and several other credit executives in other Fortune 500 firms when we would all get together to discuss credit-related research. I have a great deal of admiration for Norm and his practical, progressive, visionary approach to his work and life. In the same vein, Lamar Potts and his team in worldwide financial services at Apple provided me a global platform to implement my ideas. I owe Lamar a great deal having the belief in me to engage with me for four very productive years and for being a true friend to this day.

I have learned an unimaginable amount in my entrepreneurial efforts from so many people – colleagues, investors, and customers. There are too many to list here. One person stands a clear distance from all in this regard. Mark Nunnelly has been an extremely valuable mentor, incredibly supportive and a true friend. I have learned a tremendous amount from him both about business and life.

I owe a deep debt of gratitude to my senior team at RAGE which has believed in me for over 20 years through successive ventures and working with whom, I have been able to generate and implement so many of the ideas in this book. Aashish Mehta, Jim DeWaele, Monty Kothiwale, Nadeem Yunus, Rummana Alam, Srini Bharadwaj, you have been a bedrock of support for me and the ideas in this book. Even when it might not have made sense to you at that time, you went along enthusiastically trusting my vision. Thanks also to Joy Dasgupta and Vikram Mahidhar, both of whom have added immeasurably to the conversation surrounding this book in a very short period of time.

I am equally indebted to our wonderful team in India. While I have benefited from my interactions with all RAGE teams, I have to single out the RAGE AI Platform team – Vishaal, Nitin, Manasi, Amit J, and Atin for their passionate belief in our challenge of conventional wisdom. Vishaal and Nitin, in particular, have truly kept alive our pioneering quest to find an effective computational paradigm for natural language understanding.

This book has gained immensely from the numerous reviews of earlier drafts by Rummana Alam, Joy Dasgupta, and Vikram Mahidhar. I am most appreciative of Sanjiv Chopra's constant encouragement and reminders in our frequent meetings at Starbucks. Special thanks also to Rummana who kept nagging me to commit to writing the book and then constantly reminding me to finish it. Thanks also to Andraea DeWaele for reviewing the book for language consistency, flow, typos, and format consistency with the editorial style requirements at Wiley.

I am lucky to have such a cooperative publisher and editorial team at Wiley. Steve Quigley, Jon Gurstelle, and Allison McGinniss have been terrific to work with. They have been patient as I have kept delaying timelines amidst my compulsions running RAGE.

Above all, I am blessed with a wonderfully supportive family, my lovely wife Pratima, and our wonderful girls, Meghana and Nandini. They have borne the brunt of my constant preoccupation with intellectual and entrepreneurial pursuits with unconditional love and encouragement. I am truly thankful to them.

Over the last 28 years, I have learned from and contributed actively to the understanding and practice of knowledge-based technology and finance, first in an academic capacity and later in an entrepreneurial capacity. I have successfully created and commercialized a number of significant innovations starting with my first entrepreneurial venture, eCredit, and in subsequent ventures. My work over the previous 25+ years on knowledge process automation and more recently, tractable/traceable machine intelligence have fructified into a robust body of knowledge which I believe has great relevance in the context of the information and technology revolution that is upon us.

There are several reasons for me to write this book at this juncture of my life. First, I would like to lift the ongoing active conversation around big data and machine intelligence to a higher more strategic level by recognizing the rightful place for such intelligent technology in an enterprise architecture. Senior business executives reading the book should get a sense of how to leverage machine intelligence in their strategic and operational activities. Second, by describing real life solutions in a robust conceptual setting, I hope to afford practitioners an opportunity to extrapolate the solutions and ideas to their own situation. Third, I have seen many hype cycles come and go before the recent big data and machine intelligence hype cycle. I believe the book offers important insights that could minimize the disappointments that invariably follow a hype cycle. Firms should not think about big data in isolation. Firms can't lose sight of their existing operational issues. Firms should not blindly adopt computational statistics based machine learning without understanding the fit with the problems they are trying to solve. And finally, the book provides the opportunity for me to add to the body of knowledge in the field and hopefully enable new research and advances by others.

This book will be interesting to CEOs, CXOs, senior executives, data scientists, information technology professionals, consultants, and academics alike. I have attempted the difficult task of balancing the content so it does not get too technical and at the same time, include enough rigorous material to satisfy the more technically inclined. I hope you, the reader, find it worthwhile.

PART I

CHALLENGES OF THE DIGITAL AGE

CHAPTER 1

THE CRISIS HAS NOT GONE AWAY: OPPORTUNITY BECKONS

1.1 INTRODUCTION

It has been over twenty years since the first edition of the Champy and Hammer book *Reengineering the Corporation* (Hammer and Champy, 1993) took the business world by storm. Yet today, not a single company is satisfied with the speed with which it can respond to fast-changing market conditions. The competitive dynamics have in fact multiplied, with the Internet producing information at a stupendous rate and adding another dimension to the already complex landscape that businesses have to deal with – information. Social media has more recently added to the din, raising both the specter of uncontrolled information flow and the opportunity to reach consumers directly as never before.

The radical re-engineering era had been preceded by two other collective movements to improve competitiveness, referred to as total quality management (TQM) and Six Sigma. It was then, in the late 1980s, that, while I was designing a knowledge-based system for the Global Financial Services group at Apple, I learned of the re-engineering movement in one of my frequent visits to meet with the team at the Cupertino, California, headquarters. I was confronted by a senior executive with six massive binders of TQM flows.

The Intelligent Enterprise in the Era of Big Data, First Edition. Venkat Srinivasan.

The company had adopted TQM. I asked my client about when and how they expected to implement the quality processes in the binders. His response, "ten years and we won't be done even with a small fraction." Those binders were never implemented, and the entire company division morphed into a different form as the business underwent rapid transformation over the next couple of years.

Business process re-engineering (BPR) was but another dramatic attempt to effect corporate change. BPR was described by its proponents as a "fundamental rethinking and redesign of business processes to achieve dramatic improvements in critical measures of performance such as cost, quality, service and speed" (*The Economist*, 2009). BPR offered the tools to more effective company performance by breaking the business processes apart and identifying more efficient approaches. The analyses of these processes were to be end to end and across functional domains. The argument was that functional groups would over time become protective of their turf and withhold information in the fear that changes could lead, in some circumstances, to the elimination of their steps in the process. To be sure, lack of a big picture view of the enterprise's business process often does lead to suboptimal decisions at the function level without regard to the efficiency of the overall process. But the BPR strategy proposed an analysis whose result would be a reassembly of the business process in a more radically efficient way.

Among the several corporations that enthusiastically embraced BPR and the Six Sigma processes were Motorola and General Electric. While there is no denying that Six Sigma brought awareness for process quality and granular measurement, the act of studying, modeling, gathering, and then implementing optimal processes did take too long. It was beneficial only in business operations where there is a relative underlying stability, and that in today's fast-paced world is a significant rarity.

While a few notable successes have been reported, by and large, re-engineering did not find the kind of success the runaway popularity of the 1994 book would have implied. Hallmark reportedly completely re-engineered its new-product process, and Kodak re-engineered its black-and-white film manufacturing process to cut in half the firm's response time to new orders (Hammer and Stanton, 1999; Smith and Fingar, 2003).

A critical failing of the BPR and TQM movements at that time was the lack of sophisticated technology paradigms. Without technology paradigms being in step with the thinking behind BPR or TQM, there was no real chance of widespread success. While the BPR proponents, including Champy and Hammer, did recognize technology to be an enabler, they did not recognize technology to be the key enabling factor behind re-engineering. Their view of the new world of work revolved around the reorganization of business processes away from the silo mentality – that of one person executing a business process from end to end instead of creating a team of different people focused on different aspects of the process.

In all likelihood, systemic BPR contrarians scuttled the visionary ideas or reduced them to ordinary games. Of various other impedements, the most pervasive inhibitor was and continues to be the fragmented, inflexible, costly, nonresponsive legacy business technology landscape that underlies all the mission critical business processes that large organizations run on. In fact, it takes the best business leaders and brilliant technologists to successfully work around this encumbrance and deliver business transformation goals, despite being constrained by technology that was only in the recent past considered game changing.

The case studies that follow are based on real experiences. They illustrate how this challenge upturned the launching of new businesses, revenue generation of existing businesses, the customer experience, and operational efficiency.

CASE STUDY 1: CLIENT ONBOARDING IN HEALTHCARE SERVICES

One of the largest third-party administrator of healthcare benefits was struggling to scale to address the variability in their client systems and in efficiently onboarding them

- Client onboarding time at one of the largest third-party administrators for payroll and benefits was greater than 90 days.
- Onboarding of new clients was not possible for approximately three months in a year, while existing clients had to go through their annual renewal.
- The process had been studied repeatedly by every new business leader, re-engineered, partially offshored over the years, new IT systems introduced, but the 90-day barrier remained unbreachable.
- Onboarding every client was a software development project; same was true for any change in existing clients.

- The resulting 25% annual revenue loss from new customers and unacceptable customer experience for existing ones, was a constant source of significant frustration in the business.
- The business was losing customers routinely.

Most Viable Alternative Continue to target process improvement via Six Sigma projects and identify productivity improvements. Accept the 25% revenue abandonment rate.

CASE STUDY 2: A GLOBAL FINANCIAL SERVICES COMPANY'S NEW PRODUCT LAUNCH FOR ITS SMB CUSTOMERS

The new product launch was a key part of its future growth strategy and time sensitive.

- The company invested significantly in demand generation activity, leading up to the launch.
- This was a transaction heavy loan-origination business requiring an enterprise-grade fully automated platform.
- Six months into the start-up, the business found itself stalled even before the launch. The underlying process-technology framework was taking too long to get production ready.
- Pre-launch market signals pointed to significant adjustments to product specs. The packaged application the organization was counting on was already customized beyond the breaking point and resembled a legacy application before the first day in production.

Path Forward Shut down the project or secure additional start-up funding, inform the market of a 18-month delay, come up with a set of clear requirements (that would remain stable over this 18-month period) and engage a new technology partner.

CASE STUDY 3: LARGEST WORLDWIDE WEALTH MANAGEMENT FIRM LOOKING TO TRANSFORM ITS CUSTOMER EXPERIENCE AND FINANCIAL ADVISOR SERVICES BY REVAMPING END TO END PERFORMANCE REPORTING

The firm needed to upgrade its customer experience in order to retain its valued customers.

- The firm's packaged performance reporting and analytics tools, augmented by internally developed point solutions, were reliable and robust but inflexible and outdated.
- Two outsourced vendors who received portfolio statements from multiple custodians manually entered data from paper statements often over 500 pages long.
- The outsourced workforce grew in tandem as the firm offered broader multi-custodian reporting services to its customers.
- Reports from multiple custodians were generated once a quarter and were available 45 days after quarter end, whereby transactions made during a quarter were reported anywhere between 45 and 135 days late, rendering them useless.
- Regulatory exposure from minimally governed outsourced processes and the loss of mission-critical undocumented tribal knowledge with unplanned attrition was on the rise.

Path Forward Keep operations cost flat by moving resources to a low-cost location, improve cycle time by staffing up, manage process quality and regulatory risks by implementing high touch governance processes. This was a non-scalable arrangement. The cost of operations or a per-transaction basis was rising as growth in errors outpaced any benefits of scale. Or offer multi-custodian reporting to fewer clients which would negate a significant competitive advantage the firm had.

CASE STUDY 4: MARKET CHANGING ACQUISITION BY A GLOBAL CORPORATION

As is the norm, the deal value assumed a seamless and successful integration of the acquired firm and the delivery of all the resultant synergies.

- The acquiring firm had 800 major IT applications and countless point solutions laid out in several sedimentary layers of technology over forty years or more.
- During this time, each successive CIO inherited an ever more complex environment with the mission to standardize, simplify, and establish IT as a strategic asset for the company, but in the end tying the applications ecosystem in ever tighter knots, due to forces beyond their control. Off the shelf solutions had become upgrade proof.
- The acquired company was much smaller in size, with 150 major applications. The migration plan was bookended by converged books of accounts on day-1 and an integrated procure-to-pay process by day-365.

Path Forward The integration plans from the M&A team with limited input from technology, need to be re-planned from the ground up. Cursory analysis by the acquiring company's experienced technology team showed that dispositioning the 150 major applications in the acquired company (whether they are retained as-is, retained-but-modified, migrated or shutdown) would take at least five years if there was sustained, strong, and cascading sponsorship (staffing, funding) over that period, even as other priorities emerged and leadership changed. Neither company had ever experienced anything remotely close to that. With technology being targeted with sizable post-acquisition synergy goals in the very first year, even the most enthusiastic team members were skeptical.

The default path forward shown for each of the businesses was possibly the best choice under the circumstances. Beyond these illustrative examples, most business transformation initiatives are gated by, paced by, and inhibited by their legacy technology applications infrastructure, despite having the most current business process management [BPM] and other platforms and tools.

1.2 CHALLENGES WITH CURRENT TECHNOLOGY PARADIGMS: CHRONIC ISSUES OF TIME TO MARKET AND FLEXIBILITY

The four real life examples in the previous section amply illustrate the chronic challenges of time to market and flexibility in business technology. Large enterprises abound with executives who are frustrated with the inability of technology to keep up with their demands. Software applications developed 20 and 30 years ago still power much of our business world, even as most things around them have been replaced, renewed, and reinvented, including the very hardware they run on. Unfortunately, few large corporations have the luxury of undoing the past and starting with a clean technology slate. Plans for new businesses and plans for ambitious transformation need to ride on preexisting legacy application infrastructures that come pre-loaded with the following challenges:

- Outdated legacy applications on obsolete technology running mission-critical business processes, where the coders/developers are no longer available.
- Multiple data and technology standards. An exceptionally large tools and applications inventory and a resultant suboptimal vendor base.
- Off-the-shelf packaged solutions that have been customized beyond their intended purpose and are now the new legacy applications, upgrade-proof and several versions behind.
- In-house point solutions that were developed with great intention that simply added one more point of failure in an already fragile environment.
- IT debris from previous acquisitions that were never fully integrated.
- A business that is demanding more and more responsiveness, and IT is getting disenfranchised.
- IT costs are disproportionately high.

Our business technology landscape still plays out like a black-and-white movie. We are far from having responsive business applications. Nonresponsive applications cannot serve a high-entropy environment. Why is this so? The challenge lies in the inability of the current technology paradigms to meet the demands for rapid deployment of technology to operationalize business actions or strategies.

So what are the big hurdles that have to be addressed to enable a much more responsive business architecture. Business technology has to address four challenges to enable an effective enterprise architecture – reliability, flexibility, time to market, and knowledge.

Over the last three or four decades, reliability has been largely solved. Today, if detailed specifications are provided to a qualified technology team, the team will deliver reliable applications. There are plenty of qualified programmers and architects who can deliver a fixed set of specifications. With the emergence of standards like Capability Maturity Model [CMM] and the availability of robust databases, open source software stacks have been crucial enablers in achieving reliability. However, flexibility to meet changing business needs and time to market remain significant challenges.

No business specification is ever complete. No experts, however good they are, can express all of their expertise in a few sittings. Expert knowledge can be partitioned into easily retrievable knowledge and hard to retrieve knowledge that surfaces within the right context. Building in flexibility for what you don't know is a significant challenge in the current technology paradigms.

Most enterprise applications take anywhere between 12 and 24 months from start to initial deployment, and often within limited scope. Such a long time cycle often renders the new functionality obsolete even before the new application is ready. Time to market is a huge challenge, as speed of deployment is inversely proportional to flexibility. If you want significant flexibility, it takes more time than you might think.

Finally, knowledge is a relatively new dimension. It refers to the ability to proactively analyze the enormous amount of structured and unstructured data at our disposal and provide insights for proactive action.

Why does it take 12 to 24 months to build and deploy a robust enterprise class application? The root cause lies in the methods that we use to develop applications. Any application technology stack has three segments – application development (AD) technology, methodology, and infrastructure. Infrastructure has made tremendous progress and has led the other two dimensions. The emergence of the "cloud" and the ability to provision infrastructure on demand has more or less eliminated infrastructure as a bottleneck. But in the present book we will disregard the infrastructure dimension.

While today's application development technologies are a significant improvement over technologies even 5 to 10 years ago, they have made little impact on the overall elapsed time for applications. Many technology frameworks have emerged, such as Java, J2EE, Spring/Hibernate, .Net., Java Script, and a large number of open source tools. These have made programming easier and often faster. But they have not had a huge impact on the elapsed cycle due to the basic nature of the methodology at play. In a conventional software development lifecycle, programming is in fact a non–value-added translation step in the software development process.

1.3 THE EMERGENCE OF PACKAGED APPLICATIONS

There has also been a major shift in the AD landscape since the 1990s, with mostly commercial packaged software gaining traction, such as ERP applications like SAP, and Oracle. Such widespread use of packaged applications was largely due to the length of time it takes to build in-house applications and the rapid changes in the technology landscape that the in-house organizations were unable to keep pace with. However, there have also been huge gaps between such applications and the way enterprises run their operations. Most of the gaps have had to do with the lack of business process orientation in these applications. Therefore the software does not provide the flexibility to be used out of the box and has required significant and costly customization. Moreover, customizations require extensive programming and often take an inordinate amount of time.

Business process management (BPM) software has emerged only over the last decade to fill the void between packaged applications and the business process needs of an enterprise. Initially, BPM suites provided exception handling and workflow functionality. Even today, they offer very coarse-grained support for enterprise business processes. Since most of these suites have not been designed to support a process from the ground up and end to end, it will be a while before they achieve the fine-grained flexibility to address the dynamic nature of business needs.

While there are a large number of BPM software providers, most have not fundamentally addressed the flexibility and time to market issues. BPM software is not inexpensive, nor is it rapid or flexible, unless it happens to fit a pre-packaged solution offered by the BPM software provider, and rarely is the fit so tight. Invariably, given the coarse-grained support for business processes by all the current BPMS, implementation typically involves a high amount of conventional programming. Of course, the result is a long implementation cycle and inflexibility.

The shift to packaged applications has so far come at a cost. It has taken away the ability of the enterprise to compete. This is especially acute in financial services where the business process is the product. This explains why financial services firms continue to build a relatively high level of applications in-house. Packaged applications for basic accounting and other record keeping functions that are static with no opportunity for creating competitive advantage are quite understandable. However, packaged applications become legacy applications the moment they are put into place, unless they offer significant flexibility to be tailored to the needs of the enterprise. Thus, even though, today, applications are becoming more flexibile, for the most part they provide limited flexibility to their clients.

The shift to packaged applications may be an admission of defeat by business executives and technologists due to their inability to deliver technology in step with business needs. While it may make sense to buy packaged solutions instead of hiring an expert to build the software from scratch, the ultimate goal should be to be able to tailor or configure such a packaged solution to conform to changes in business needs.

1.4 THE NEW FRONT: INFORMATION; BIG DATA IS NOT NEW; WHAT IS NEW IS UNSTRUCTURED INFORMATION

The immense untapped strategic and operational potential buried in the body of information and knowledge in cyberspace is the latest new dimension that will influence and shape enterprise architecture in the future. It offers tantalizing possibilities of a real time intelligent enterprise architecture. The technology to unlock the potential of big data and make it practical and applicable to drive new value for old (and new) businesses is still in its infancy, and lags the hype around big data by a great distance.

Predictive analytics (statistically analyzing the past to project the future), has been around for decades and will become even more powerful as it is pointed at bigger and bigger, but structured, data. Social media analytics, a more recent phenomenon, has fired the imagination and investments of organizations looking to get a handle on general public opinion toward their product, brand, or company based on chatter on social media platforms, and then using the same media to steer opinion or take other action. And there is a rapidly growing population of data scientists, who are being given the task of taming big data by applying mathematics, backed by organization muscle.

What continues to be out of reach of predictive analytics, rudimentary social media listening, and lightly armed data scientists is the vast and exponentially growing universe of *unstructured data*, both outside and within corporate walls.

As noted in IDC's Digital Universe Study of June 2011, sponsored by EMC Inc., 90% of all data in 2010 was unstructured and, given the technology at that time, mostly inaccessible to machine processing. The big data technology industry is growing at a rapid clip of 40 to 60%, depending on which

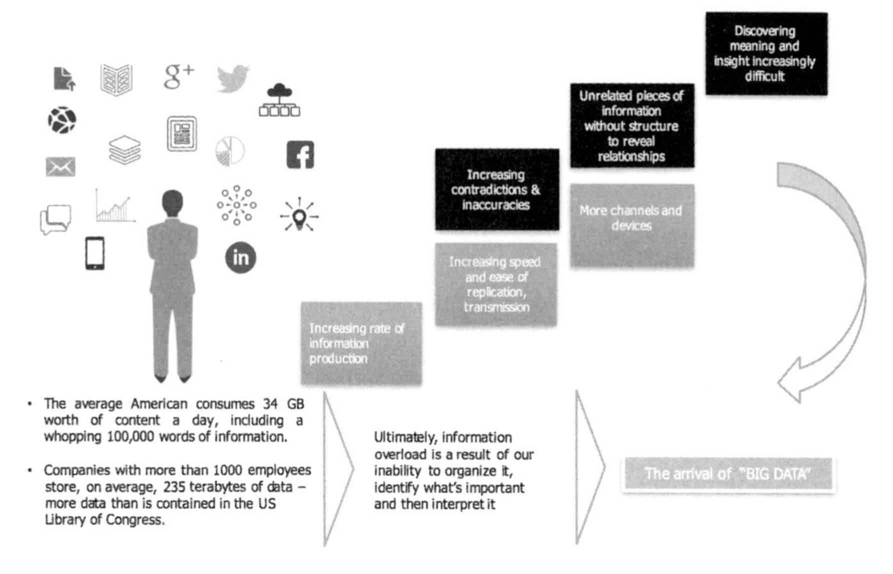

Figure 1.1 Information overload problem

source you trust. However, more than 80% of this spend is directed toward infrastructure, storage, and databases. Only 20% of this investment is directed toward big data business applications (wikibon, 2012). Of the 20%, only a small portion is focused on tackling the biggest big data opportunity – analyzing unstructured data.

The fact that we are producing information at a rate that has long surpassed human processing capacity has even led us to apply words and phrases like "noise," "chatter," and "information overload" (see Figure 1.1) to describe this low signal-to-noise ratio phenomenon, as though the problem was somehow inherent in the information and not in our ability to deal with it.

Most enterprises are chomping at the bit to access and leverage the insights from such vast amounts of unstructured data.

While the rush to big data has attracted the best minds and big money, staying on top of and discovering meaning in an ocean of unrelated fragments of information, emerging incessantly, and available 24/7 at ever growing volumes, has turned out to be mostly a losing battle for the pioneers. Leading industry analysts suggest that the big data hype and promise has, for the most part, given way to a less ambitious inward focus in major corporations for the foreseeable future. The mining for insights within internal structured data has left the business of tapping into the potential of the world of unstructured information "out there" for another day.

1.5 ENTERPRISE ARCHITECTURE: CURRENT STATE AND IMPLICATIONS

Fundamental to the questions of enterprise competitiveness is the question of how enterprises organize themselves and compete? Such an enterprise architecture is central to the issues outlined above and has the potential to undergo a profound transformation as the solutions to the above-mentioned issues emerge (Figure 1.2).

Post Adam Smith, enterprise architecture was based on the principle of division of labor and centered around specialized functions. Each function was supposed to become highly competent and prevent skill fragmentation in producing the end product at scale.

In the modern era with a significantly flat world and with a pervasive technology influence, there have been many proponents of a business process centric architecture. Hammer and Champy forcefully articulated this position in their 1993 seminal work.

There has been a lot of discussion on whether the enterprise should be organized along functions or by process. All these structures nevertheless imply that the architecture is "people led." People are at the heart of the organization and execution of the work. We can, in this regard, think of an enterprise architecture as consisting of design/strategy and execution. Business

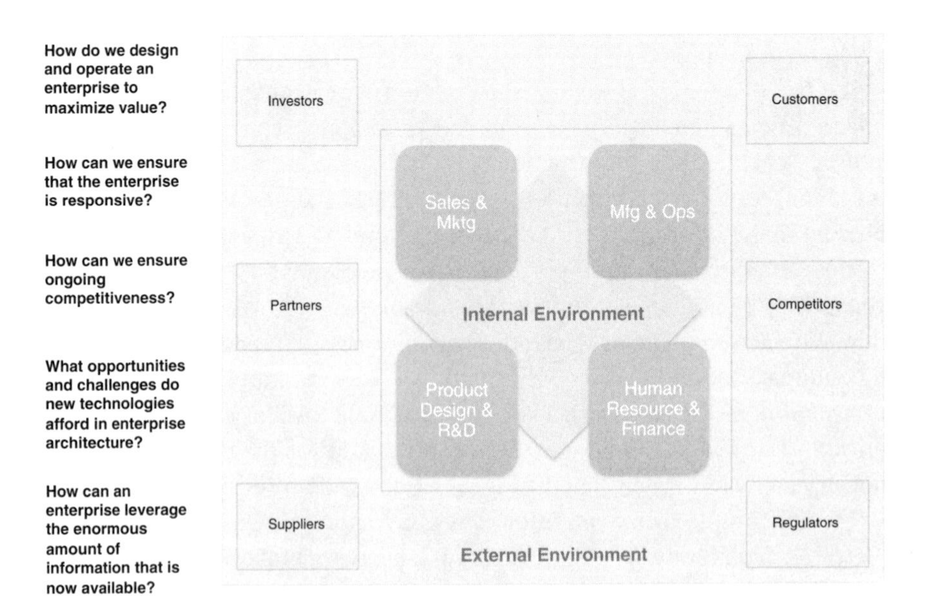

Figure 1.2 Enterprise environment

processes are designed by humans with the necessary expertise and reflect their vision, knowledge, and the overall mission of the organization. Once all components are designed and put in place, then all relevant business transactions constitute a repeatedly executed business process.

1.6 THE INTELLIGENT ENTERPRISE OF TOMORROW

Technological advances, the explosive growth in information, and the emergence of instant, viral communication have combined to create a dynamically changing environment for enterprises. This has had far-reaching implications for how an enterprise architects itself. We believe that, for enterprises to be successful in such an environment, they need to rethink every aspect of their business from the ground up.

Indeed, it may be that the fundamental "people-led" approach to enterprise architecture should change to allow machines to take on a larger role in the enterprise architecture. How much of a role can machines have in the architecture? To be sure, this will vary across industries and firms, but there is no doubt that machines will increasingly assume more of the execution role. And they will increasingly assist humans in the design aspect with assembled intelligence that humans simply cannot collect on their own. Machines will nevertheless need to be instructed by human experts in assembling such intelligence and such instructions will need to be continually monitored and adjusted.

These are the challenges explored in this book. In the next six chapters we attempt to outline what form an optimal architecture for the intelligent enterprise would take.

REFERENCES

Hammer, M., and Champy, J. (1993). *Reengineering the Corporation: A Manifesto for Business Revolution*. New York: HarperBusiness; revised updated ed., HarperCollins, 2004.

Davenport, T. H. (2005). The coming commoditization of processes. *Harvard Business Review* (June).

Hammer, M., and Stanton, S. (1999). How process enterprises really work. *Harvard Business Review* (November–December): 2–11.

Gantz, J., and Reinsel, D. (2012). *The Digital Universe in 2020: Big Data, Bigger Digital Shadows, and Biggest Growth in the Far East*. Framingham, MA: IDC.

Economist, The. (2009). Business process reengineering, February 16.

Smith, H. and Fingar, P. (2003). *Business Process Management: The Third Wave.* Tampa: Meghan-Kiffer Press.

Wikibon.org (2012). Wikibon_IT_Transformation_Survey: 2012 is the year of the cloud, 2012. http://wikibon.org/wiki/v/Wikibon_IT_Transformation_Survey:_2012_is_The_Year_of_the_Cloud.

PART II

AN ARCHITECTURE FOR THE INTELLIGENT ENTERPRISE

CHAPTER 2

EFFICIENCY AND AGILITY

2.1 INTRODUCTION

Most enterprises strive to become more efficient and agile in reacting to the changing demands of their customers, and the changing dynamics of the business environment. However, in today's connected world, the real key to achieving efficiency and agility lies in effective leverage of technology. As was discussed in the preceding chapter, technology has so far failed to provide the efficiency and agility that most businesses expect and need.

In this chapter, we trace our steps through the evolution of a technology paradigm that can make near real time flexible software development feasible. We believe this will usher in an era of unprecedented speed and responsiveness.

2.2 THE PROCESS-ORIENTED ENTERPRISE

There is considerable discussion especially in the popular press on the merits of process-oriented organizations (Hammer, 1996; Hammer and Champy, 1993; Davenport, 1993, 2005). Most firms nowadays believe that to be

The Intelligent Enterprise in the Era of Big Data, First Edition. Venkat Srinivasan.

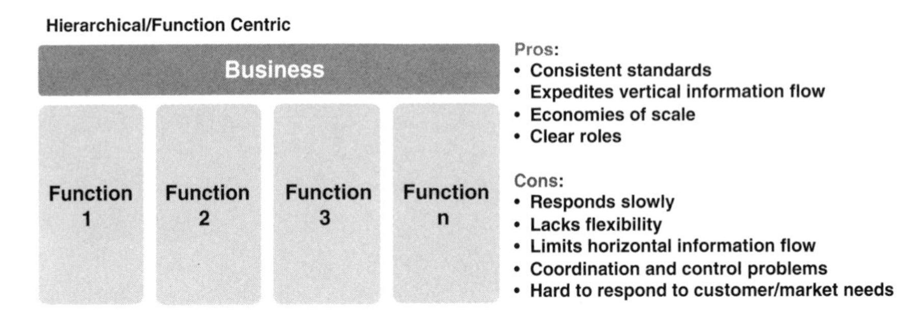

Figure 2.1 Functional enterprise architecture

competitive and efficient they need to be process oriented in terms of their organizational structure instead of organizing with a functional orientation.

Figure 2.1 illustrates a typical functional architecture. Many corporations are still organized on a functional, hierarchical basis. They are structured hierarchically within functional areas like sales, marketing, finance, or production. They reflect the importance placed on skills in an internally focused view of organizational governance. This corporate structure emerged in the post-Industrial Revolution when division of labor and specialization were advanced by big corporations to standardize their production processes and maintain a competitive economies of scale edge.

The conventional wisdom was that such an organizational structure ensured consistency in the manufacture of goods and the diffusion of standards within that function. In organizations focused on bettering their economies of scale, the cross-functional implications are dealt with at higher management levels of the organization.

However, such organizations often create functional silos within their structures. Such organizations, with few exceptions, also often create less than optimal experiences for customers. This is because there is no holistic focus in the organization on delivering exceptional customer experience. Often there are frequent breakdowns in the handoff in process execution between functional areas resulting in scrambles to protect customer experience.

While function-centric organizations did enable enterprises to scale in the post-Industrial Revolution era, the emergence of the Internet, along with advances in mobile communications and technology, have made the weaknesses of the function-centric organization readily apparent. In this flat dynamic world, the ability to adapt and innovate business processes end to end has become critical to maintaining a competitive edge.

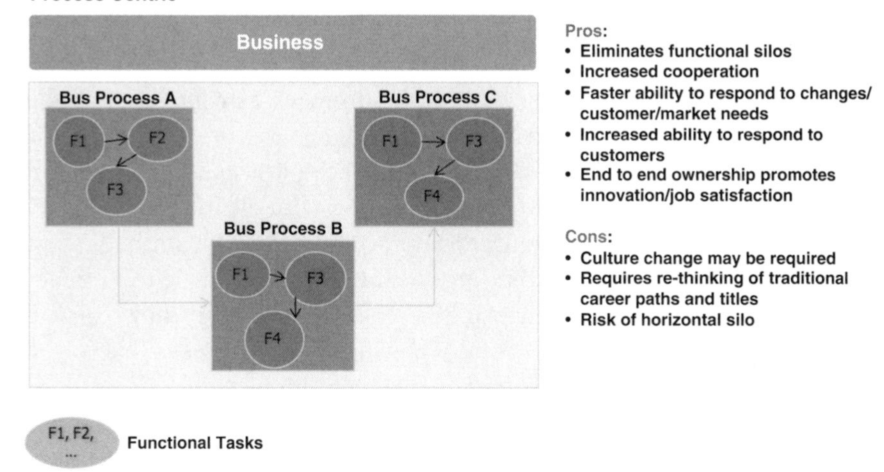

Figure 2.2 Process-centric enterprise architecture

The process-centric organization's (Figure 2.2) key feature is, in contrast, its business processes. Hammer and Champy (1993) define the business process as follows:

> ... a collection of activities that takes one or more kinds of input and creates an output that is of value to the customer.

Thomas Davenport (1993) defines a process more succinctly:

> Simply a structured, measured set of activities designed to produce a specified output for a particular customer of market. It implies a strong emphasis upon *how* work is done within an enterprise, in contrast to a product focus's emphasis on *what*. A process is thus a specific ordering of work activities across time and place, with a beginning, an end, and clearly identified inputs and outputs: a structure for action.

In Davenport's definition of process, there is a key distinction between "how work is done" with "what work is done." "Process" is more exactly delimited to "how." In both interpretations, there is a clear emphasis on final value for the customer.

Rummler and Brache (1995) broaden the definition in making a distinction between primary and support processes based on whether or not customer value is directly discernible from the process regardless of whether it is nevertheless essential.

Our definition of a business process is much more holistic. We define a business (or corporation) to be a collection of processes with corresponding goals vis-à-vis its stakeholders (customers, employees, investors, regulators, etc.). In our view, a process is a defined sequence of business activities with input and output parameters for every task and with clearly defined objectives with respect to a stakeholder. The process has to do with both "how" the tasks are to be done and "what" tasks are to be accomplished. We believe that the "what" in most instances is not separable from the "how" because often "what" specifies the objectives of the "how" in a process. This way the process definition does not lack purpose and measurement.

We do not define process synonymously with workflow, as this considerably weakens the definition of process by ignoring the "work" in the process or main business tasks of the process. Our definition of process is more inclusive and envelops both "work" and "workflow" associated with the process.

Each business process in a process-centric organization typically is owned by a "process owner." Process teams led by the process owner will include members from different skill or functional groups and have end to end view. Therefore process teams are focused on the ultimate goals of the process whether the goals are delivery of customer experience or, say, order to cash.

A process-centric organizational structure eliminates functional silos and facilitates convergence toward the process objective. There is no friction between different functional groups that execute their separate tasks in the process. They can more naturally apply their skills and continue to innovate.

Regardless of the approaches and methods they recommend on how to become process oriented, many authors see significant benefits accruing to process-oriented organizations. Hammer (2007) argues that organizations can improve the quality of their products and services, time to market, their profitability, and other dimensions by focusing on customer-centric processes. Hirzel (2008) similarly cites reductions in cycle times for product delivery, responding to customer needs as a result of adopting process orientation. Ligus (1993) gives very specific estimates of improvement in various metrics – delivery time reduction by 75–80%, reduction in inventory levels by up to 70%, and even market share gains with process orientation. Kohlbacher (2010) cites increase in customer satisfaction, improvement in quality of products and services, overall cost reduction, improved speed in responding to customer needs as the major benefits of adopting a process-oriented system. Indeed, there seems to be significant consensus that process orientation provides organization with more efficiency and agility relative to not being process oriented.

2.2.1 Becoming Process Oriented

In reality, few organizations have been able to achieve true process orientation largely because of the limitations of technology and the significant culture change required to achieve such orientation. Even those organizations that have set up and now utilize process-centric functions, struggle to effectively manage the cultural change required to run their process-oriented systems. True process orientation goes far beyond organizing by processes. It requires that the organization's members undergo a change in mindset and internalize what it means to be process-oriented. Hammer (1999) points out that the power in most companies resides in vertical units – sometimes focused on regions, sometimes on products, sometimes on functions – and these fiefdoms still jealously guard their turf, their people, and their resources. This results in an environment whereby people get pulled in different directions, raises organizational friction, and undermines organizational performance.

Also we find that ultimately extreme frustration with the inability of function-centric organizations to respond to market needs leads to major transformation projects to become process oriented. These frustrations can build up to the point whereby internal inefficiencies lead to a wholesale mandate to internally or externally transform themselves. Such projects often cut across functional lines. You often hear of projects like "one and done," which implies that many redundant points of interaction, or information entry, or customer touch, caused as a result of functional orientation, need to be reduced to one integrated point.

I one time worked with a large information products company that had 19 different systems that gathered more or less the same corporate fundamental financial data. Often there were errors and serious quality issues in what they delivered to their customers. The issues built up to the point where senior management had to sanction a transformation project that was titled, "one and done." The project had a cross-functional owner who had senior management support to cut across organizational silos to implement "one and done."

Recently, I came across a global logistics company that has many systems around the world, organized by geography, that interact in servicing customer orders. This company is not able to provide a consistent and accurate response to customer inquiries and complaints, and as a result customer satisfaction levels are low. There is no coordinated customer response process, and often the same customer operating in two different geographies gets different services based on where they log their feedback. The responses of customer service representatives range in quality based on their knowledge of the company's products and processes. The company found that often the initial classification of the customer complaint was incorrect and frequently the classification changed during the interaction with the customer. The back-and-forth with the customer caused a lot of frustration at the customer end. This company now has launched a huge transformation project. The project has one owner across geographies and functions and has the mandate to create the capability to provide customers with consistent, accurate, and timely responses.

Both the cases above were instances of issues building up to point where the organization is compelled to effect a dramatic change in order to remain functioning. Obviously, by the time the organization recognizes the need, a lot of its reputation has soured.

There is considerable literature on measuring the level of process orientation that a business has achieved and its level of process maturity. Researchers and practitioners have proposed a number of business process maturity models – the business process orientation maturity model (McCormack and Johnson, 2001), the BPM capability framework (Rosemann and de Bruin, 2005), the process and enterprise maturity model (Hammer, 2007), and the object management group business process maturity model (OMG, 2008) are among the commonly referred to models. Tarhan et al. (2015), however, find that, after a comprehensive survey of the literature, there is a scarcity of evidence linking the level of process maturity of an organization, as measured using one of the above maturity models, to improved business performance. While this appears to be somewhat contradictory to the findings on the benefits from process orientation cited earlier, the difference perhaps lies in the degree of benefit and the definition of the business process maturity models.

In practice, most organizations are a hybrid of process- and function-oriented approaches. Even those organizations that have made the transition to process orientation are process oriented up to a point and become hierarchical beyond a certain level.

2.2.2 Why Must We Choose?

Why do organizations have to choose between process and functional orientations? The debate on the pros and cons of process versus functional orientation inherently reflects the limitations of technology and the limited role technology has played in enabling organizations to become efficient and agile. We think the lack of effective technology paradigms is the key reason why firms face such difficulties in becoming process oriented from the ground up. If technology paradigms were flexible, then firms would have found it much easier to become process oriented.

We can even take this one step further. Why can't technology enable organizations to have dynamic structures that serve both functional and process driven needs seamlessly? With technology advances, we believe an organization can be process and functionally oriented at the same time without getting pulled in both directions. We present such a scenario later in this chapter.

2.2.3 Design and Execution

There is another dimension to enterprise organization that has not had much attention in the literature – design versus execution (Figure 2.3).

It is important to recognize the difference between design and execution in enterprise architecture. Design refers to the design of various strategies, business models, and business processes. Every business process is designed by a person and is executed by one or more people in the organization. Sometimes these are the same people. These two roles, though are very distinct, and separate in a mature organization. Leymann and Alternhuber [1994] recognize that there is a build time and a runtime aspect to business processes.

Once business processes are designed, they are executed repeatedly by the execution part of the organization. Regardless of whether an organization is function or process centric, some individuals will be focused on design and the bulk of the organization on execution.

The execution portion of any business process has naturally been aligned with automation because of the inherent repetitiveness of the assembly process. Design has been the purview of humans until recently.

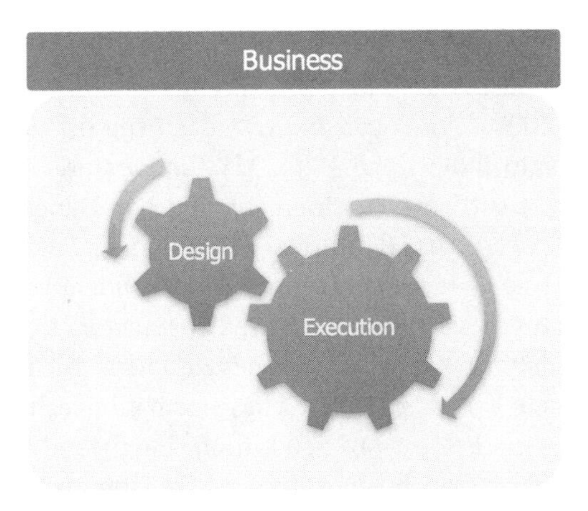

Figure 2.3 Design versus execution in enterprise architecture

Designers are also referred to as "process owners" and are often subject-matter experts or operations/process experts. They do draw on others' expertise in thinking about the best way to design a process. Executors, in contrast, are charged with executing the process defined by the designers.

Design requires a distinct set of skills, expertise, and information. Execution requires an entirely different set of skills, expertise, and information. Design is informed by the results of execution.

2.3 ROLE OF OUTSOURCING IN CREATING EFFICIENCY AND AGILITY

Offshoring is now being recognized as a societal revolution on par with the Industrial Revolution (Davis, 2004). For a long time, manufacturing was largely done in the developed world for developed markets, and services were similarly produced and consumed within each separate market. However, by the 1980s, outsourced manufacturing in offshore locations, particularly China, became fairly well established.

The most talked about and visible outsourcing has been the transfer of manufacturing processes to countries that have very low labor costs. These have caused significant disruptions in entire industries in the United States, especially, in textile manufacturing. These manufacturing operations have moved around the world, transferring to new sites as relative labor cost differentials change. Many organizations initially outsourced manufacturing to Mexico and then to countries in Asia.

The outsourcing of technology-related services emerged in the late 1980s and has continued to grow significantly in the 1990s and beyond. Covansys, based in Farmington Hills, Michigan, was one of the first to specialize in helping American businesses move offshore and set up an Indian facility in 1992 (Reingold, 2004). American Express has been offshoring a variety of business functions to India since 1994. My first venture, eCredit, went offshore with a captive software development center in Bangalore in 1994. GE Capital launched a captive, GE Capital International Services (GECIS), in India, which has since been spun off as an independent company, Genpact. All major financial services firms have used cost factors to justify offshoring many functions, ranging from software development to customer servicing. Between 1989, when Kodak outsourced its information technology, and 1995, the IT outsourcing market grew to $76 billion (Lacity and Willcocks, 1998).

Both manufacturing and business process outsourcing has been largely cost-driven arbitrage. Conceptually, offshoring and outsourcing are the realization of the adoption of the free trade doctrine and rooted in principles of producing goods and services where they can be done most efficiently. Sawhney (2002) termed this leverage of relative cost differentials across countries as "global arbitrage." Moxon (1995) empirically examined inexpensive labor as the primary motivation for offshore production for the electronics industry. The importance of production costs has also been suggested in information system studies (Ang and Straub, 1998; Wang, 2002). Clearly, advancements in communication and computer technologies in recent years have made this argument ever more plausible, and offshoring opportunities ever more feasible, particularly in the service domain. With the increase in the "export" of high-skilled, highly paid jobs, the expectation for cost advantages has also increased. While there are other supporting arguments in favor of outsourcing like quality and delivery time, there is no large scale evidence supporting these arguments as the primary cause.

While the number of offshoring destinations has increased over the years, with China, Malaysia, the Philippines, and South Africa rapidly encroaching on India (*The Economist*, 2003a,b), India still remains the world's topmost destination for business process outsourcing. Using the Indian BPO industry data as a proxy for the global BPO industry, from a heady growth of over 60% in 2003 to 2004, growth rates in the BPO and ITES industries in India have dropped significantly to low double digits in 2014 to 2015 because of the following factors:

- *Wage inflation* India's cost advantage as an offshoring destination, its trump card, has dropped by 30–40%. And it is dropping further every year.
- *High attrition* At 40–50% attrition, BPOs have the highest attrition rate of any industry and companies have to hire half their workforce every year.
- *Increased regulations and controls* Outsourcing of financial market oversight jobs by global banks to India has come under the scanner several times. For example, outsourcing of key oversight jobs by global banks (British giants HSBC and Standard Chartered) to India has come under the regulatory scanner in the host countries for ineffective controls against suspicious financial transactions.

Work done by a different set of people in a different location is not going to result in any process transformation in and of itself. Incremental Six Sigma initiatives won't do much for a process already running at 99.5% of Service Level Agreements [SLAs] constructed under the premise of a manual process.

From the perspective of outsourcers, pure outsourcing without significant re-engineering of the processes will not make businesses more efficient and agile. After a successful wave of cost arbitrage driven outsourcing, the next wave will have to be led by quality and time to market advantages. This calls for technology leverage and expertise. Outsourcers are beginning to move from a pure FTE-based outsourcing model to managed services. More important, they are insisting on understanding and seeing how BPO firms will deploy technology and automation to improve their outsourced processes. A perfect storm is in progress, with customers seeking to cut additional backoffice costs due to continued budget pressure, while suppliers are trying to create additional services and the revenues that go with them (*ComputerWeekly*, 2015). In an analysis of India's offshoring industry, McKinsey analysis concludes, "Innovation will be the key to maintaining and even expanding their market share. Business models that continue to focus on low labor costs won't suffice." It finds that in the post-recession environment, there will be constant pressure to lower costs and improve services (McKinsey, August 2009).

As I had noted in an interview to the *Wharton Knowledge* magazine in 2009 (Srinivasan, 2009), I have never believed in the long-term viability of a pure "lift and shift" approach that has dominated the BPO industry and still does. While the one time cost savings from cost arbitrage made sense in the early days, and perhaps still does in some cases as a quick first stop, the idea of simply shifting the processes to another location, without improving or optimizing it where necessary, does not make sense in the long run. It has served both sides very well for a couple of decades, but now the time has come for a fundamental reshaping of the BPO and ITES industries. Recent advances in technology provide an opportunity and challenge to reinvent how technology is leveraged for efficiency and agility. Whether BPO and information technology enabled services [ITES] firms are able to exploit it to their advantage remains to be seen.

There is growing consensus now that margins in the outsourcing industry are headed downward. For more than a decade, outsourcing firms enjoyed one of the highest profitability levels for any industry, but a combination of forces is pushing margins to levels more typical of commoditized products and services (Nandakumar and Prasad, 2013). Many BPO firms are reinventing themselves, shifting their focus finally to automation and increased technology leverage.

Genpact has introduced its Lean Digital^SM approach. Genpact looks to enable large companies to re-architect their middle- and backoffice operations and achieve measurable impact such as growth, cost efficiency, and business agility. "Lean Digital ushers in a new era in how our clients are approaching digital transformation – a shift that has driven adoption of these methods in many of our banking, insurance, manufacturing, life sciences, and consumer product clients, and that has driven our investments including the creation of an innovation lab in Silicon Valley to prototype and experiment with them," said Tiger Tyagarajan, Genpact president and CEO (Genpact, 2015). Genpact recently announced a strategic partnership with RAGE Frameworks to combine artificial intelligence and its Lean Digital^{SM} approach to push the boundaries of automation in financial institutions. Similarly, Wipro expects its investments in automation, artificial intelligence (AI) and digital technology to improve efficiency and bring about a reduction of 30% in its headcount in the next three years (ET Bureau, 2015).

Recently, use of robotic process automation (RPA) software solutions that replicate manually operated processes has been in vogue in BPO firms. This has gained popularity because a large portion of BPO processes are very mechanical and involve keying in data into client systems. While robots can address such mechanical processes, robotic automation is hardly the long-term solution to increase agility and flexibility at a more fundamental business process level.

Davenport (2005) predicts that a broad set of process standards will emerge over the years, and these will provide the metrics and transparency necessary to judge whether outsourcing will be beneficial and make it easier to compare service providers. This line of discourse does not factor the transformative impact of recent technology. In fact, it assumes that technology will continue to be unresponsive to business needs. While standards-driven software development has made it easier to create reliable software, it has not changed the speed with which software is developed and its responsiveness to business needs. In the decade to come, the process of software development itself will change dramatically, resulting in an about turn in companies' approach to software development and, more important, what they outsource and how.

We believe that transformative technology will enable a company to intelligently decide which processes, if any, should be outsourced and to seamlessly integrate outsourced processes with the rest of the company's processes. The flexibility with which technology will facilitate business process execution could allow companies to own even non-core processes if they perceive some benefit in doing so. Such transformative technology will usher in a new era of innovation in all processes, core and non-core.

2.4 ROLE OF TECHNOLOGY IN EFFICIENCY AND AGILITY

Technology is clearly central to the issues of efficiency and agility. Historically, there has been considerable recognition of the role technology should play in enabling enterprises to be competitive. In the last decade and more, this has only become more so with the tremendous growth of the Internet and the interconnectedness of businesses around the world.

2.4.1 Current Challenges with Technology

Technology has long held out the promise of facilitating more efficient and flexible ways of doing business and enabling competitive differentiation. Historically, also, any large enterprise abounds with business executives who are deeply frustrated with the inability of technology to keep up with their demands. Technologists blame the application development (AD) technologies even though AD technologies have evolved considerably since the early days of computing. We have seen major evolutionary shifts in programming environments, architectures, and methodologies as a result of the knowledge gained with each phase of evolution. However, while today's AD technologies are clearly a significant improvement over technologies 5 to 10 years ago, several fundamental issues such as *time to market*, the need to cope with the *rapid changes in technology environments and standards*, and the *availability of skilled resources remain.*

There has also been a major shift in the AD landscape since the 1990s, with many packaged applications gaining significant traction. However, these applications have required significant customization and have not provided the flexibility to be used out of the box. Customizations require extensive programming and often take an inordinate amount of time. There have also been significant gaps between such applications and the way enterprises run their business.

The gap in packaged applications and business needs is largely because these applications were not designed to be process-centric from the ground up. In most cases, businesses chose to adapt their processes to the way the applications are built or to customize the applications, rendering the promise of future upgrades invalid. This is how they end up creating a major maintenance burden for the future. Nevertheless, most of these packaged applications have not provided enterprises with the efficiency and agility needed nowadays.

2.4.2 BPM Software

Business process management (BPM) software has emerged over the last decade and more to fill the void between packaged applications and the business process needs of an enterprise. Initially, BPM suites provided exception handling and workflow functionality. As the name suggests, BPM has been defined as the management of design and control of business processes (Leymann and Altenhuber, 1994). They distinguish between design time and runtime aspects of a business process. Reijers (2003) defines BPM as the field of designing and controlling business processes.

BPM platform based case management frameworks (CMFs) have been introduced by traditional BPM providers to reduce the time to market in creating case-driven solutions. CMFs are out of the box design patterns to create a unique custom solution. CMFs have been advocated where packaged solutions are not available or considered not suitable due to requirements gaps, lack of flexibility, and the like (Gartner, 2015).

CMFs based on the BPM platform differ from traditional reusable code libraries or packaged commercial applications as they are generally model driven and rely on an orchestration engine to drive execution. Many CMFs execute the solution directly from metadata and not from code generated from metadata. Some providers have built domain specific vertical frameworks on top of CMFs like call center operations.

While there are a large number of BPM software providers offering CMFs, they are, however, only capable of supporting the case workflow. Since most of these suites have not been designed with a process orientation from the ground up, it will be a while before they are able to achieve fine-grained flexibility to address the dynamic nature of business needs. Because of this, none have been able to effectively address the basic issues of flexibility and time to market in the associated application development efforts. Any project based on BPM platforms is still not inexpensive, rapid, or flexible unless it happens to fit a pre-packaged solution offered by the BPM software provider, and very rarely is the fit tight. Invariably, given the coarse-grained support for business process by all the current BPM platforms, implementation of the full project typically involves a high amount of conventional programming. Of course, the result is a long implementation cycle and inflexibility.

In a recent example, where a global financial institution needed an end to end loan processing application for a new line of loan products, estimates for the initial implementation by leading IT service providers using market leading BPM platforms including ones with CMFs ranged from 18 to 24 months. An analysis of requirements reveals that BPM providers with their current coarse-grained support for business processes would have supported only between 10–15% of the total system functionality (Figure 2.4). The rest would have had to be developed using contemporary software development methods. In this example of a greenfield application, the time to market challenges even with using a BPMS were a result of lack of functionality that can

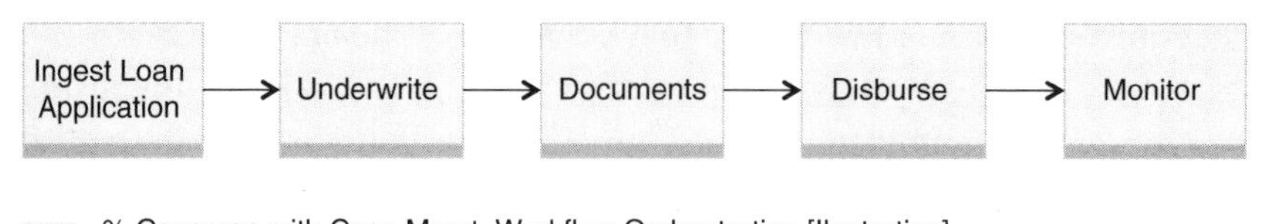

Figure 2.4 Typical BPM role in E2E loan processing application

support the full spectrum of needs for this application. Added to that is the lack of process thinking in the design of such BPMS.

Obviously, with support for only 10–15% of the application, the project could not have gained any significant time to market or flexibility benefits.

Another important point to remember is that BPMs do not impact the requirement analysis or testing stages of the software development cycle. Both of these are very time-consuming for the project as a whole regardless of the methodology used, as we discuss next.

2.4.3 Role of Methodology

Software development lifecycle (SDLC) has evolved from an idiosyncratic art to an engineering discipline. The days of learning the art from a master craftsman have given away to structure, standards, and the disciplines akin to engineering science. A conventional software development lifecycle is depicted in Figure 2.5.

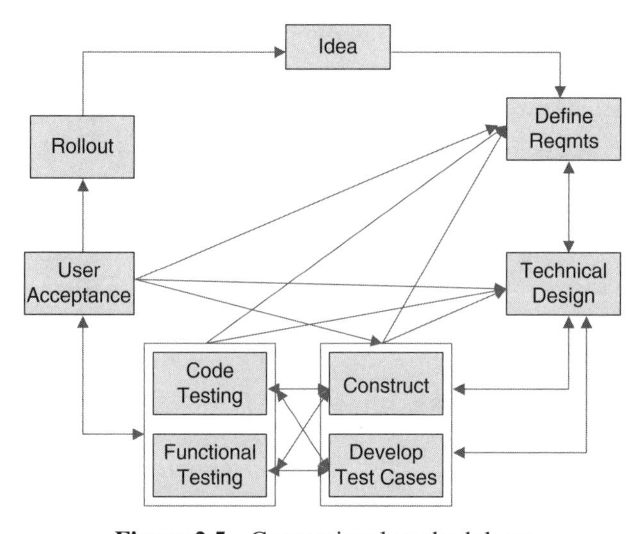

Figure 2.5 Conventional methodology

An SDLC consists of many stages as an idea gets converted into a software application. At each stage, output from the previous stage is translated to serve the purpose and audience of the current stage. These intermediate translations are necessary so that ultimately we can produce a translation capable of being acted on by a computer. With each translation, we struggle to preserve full information and knowledge from the previous stage. Obviously, each stage adds precious time to the overall lifecycle. Some of the stages require specialized skills like programming.

Over the years, there has been a substantial amount of effort in making the SDLC effective and predictable. Such efforts can be classified into two major categories:

- Methodological approaches aimed at improving the communication and translation between the various stages of the lifecycle by standardizing them and at reducing the redundancies across application development efforts.
- Automation approaches aimed at automating one or more stages of the cycle.

The emphasis on methodological approaches has raised the SDLC to an engineering science and increased the predictability and reliability of software development, in general. However, the methodologies have largely not addressed the time to market and flexibility issues. This is because they have assumed the forward engineering oriented SDLC as given.

2.4.4 Agile Not Equal to Agility

More recently, Agile methodologies appear to be gaining adoption. All Agile methodologies, loosely speaking, are shorter versions of the above contemporary methodology described in Figure 2.6. They have largely resulted from the realization that requirements change frequently and are not often precise. To address such variability in the process and to reduce the costs of discovering gaps substantially downstream from the current step, Agile methods advocate shorter iterations and emphasize less focus on static steps like documentation.

As Figure 2.6 illustrates by a generic Agile methodology, frequent iterations are at the heart of all Agile methodologies. Agile projects start with a release planning phrase followed by several iterations, each of which concludes with user acceptance testing. When the product meets minimally viable set of functionality, it is released. Users write “user stories” to describe the desired functionality. Such stories form the requirements used to estimate

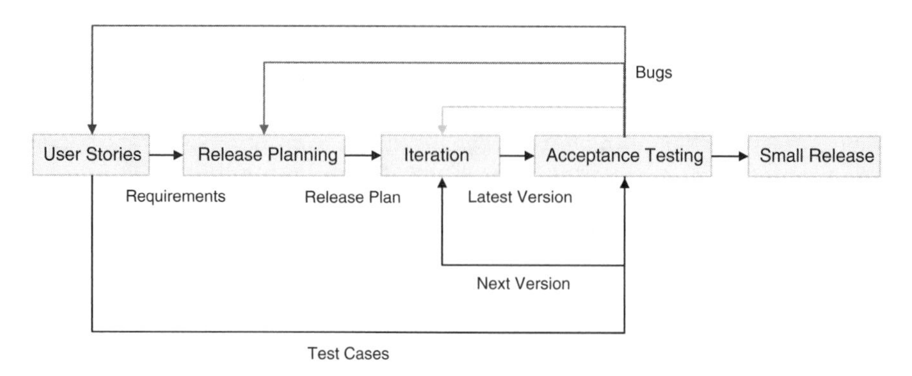

Figure 2.6 Generic Agile methodology

project execution. Execution is divided into as many iterations as necessary. Users are allowed to add more detail to their stories even when iterations are in progress. The incremental changes are accommodated in the next iteration. At the end of each iteration, users test the application and bugs are simply made part of the next iteration. The users can decide at any time to release the software if enough functionality is available.

As is obvious, this is a shorter version of the contemporary methodology defined in Figure 2.1. Important philosophical differences are the recognition that requirements are not static, and that some steps like complete requirements documentation are unnecessary and drag the process down. Agile's success is dependent on having the requirements expert on hand all the time. This is because user stories are envisioned to be very brief. The methodology relies on the expert more and less on documentation.

Agile methodologies, despite what might be construed from the name, do not result in any significant reduction in time to market or flexibility. In fact, as with anything new, we see significant misunderstanding of what Agile methodologies will accomplish. Agile methodologies should increase the probability of acceptance by the users significantly, since the users will be seeing small pieces of the application frequently and will have the opportunity to make modifications all along the way. If there is a real SME at the helm, then it is likely that there will be a minor reduction in time to market compared to a conventional SDLC. More often than not, we predict that Agile methods will increase the time to develop an application. Remember, there are several non-value-added steps like testing that will have to be done more repetitiously compared to a non-Agile environment.

So Agile does not equal agility. This is a critical conclusion that enterprises need to internalize. And it does not address the basic issues of *time to market* and *flexibility*. It is realistically only likely to impact *reliability*.

Despite significant focus on improving the software development lifecycle (SDLC), none of the approaches have made a significant difference in the age-old issues of time to market and flexibility. More important, the rate of failure or huge cost overruns of software projects is still unacceptably high.

2.5 A NEW TECHNOLOGY PARADIGM FOR EFFICIENCY AND AGILITY

There is significant evidence about the benefits of process orientation compared to purely functional organizations. Despite this, businesses find it hard to become process oriented in practice. We have also observed that technology can play a crucial role in enabling enterprises to become process oriented without going through a lot of change management.

At this point, we will outline a technology platform and methodology that can address the chronic time to market and flexibility issues and enable businesses to become both efficient and agile. We discuss a meta model-driven business process automation platform that can enable near real time software development largely without programming and purely using metadata.

2.5.1 Technology and the Process-Oriented Architecture

We begin by outlining a high-level conceptual relationship between a process-oriented business and technology. The relationship is shown in Figure 2.7.

Figure 2.7 outlines a hypothetical business. It is conceptualized as a collection of business processes, which is how it ought to be viewed [BP_1, BP_2, BP_3, …, BP_m]. We can further think of each business process as a collection of business tasks. A business process is a set of interrelated tasks that are performed in some order by a combination of humans and machines [T_1, T_2, …, T_n].

The way software applications are built is to program these tasks using a programming language or framework. What if we were to eliminate programming through an extensive abstraction layer?

A simple way to relate to this is to think of each task as corresponding to a technology component. At an abstract level, many of these tasks are common to all applications. In every software development exercise, we keep doing the same thing again and again. Every application needs technology support for similar tasks and functions – workflow, case management, interfaces to other internal and external systems, decisioning, computations, dynamic document creation, logic driven questionnaires, and so on. We can think of many such tasks that are domain agnostic. We can, of course, also think of

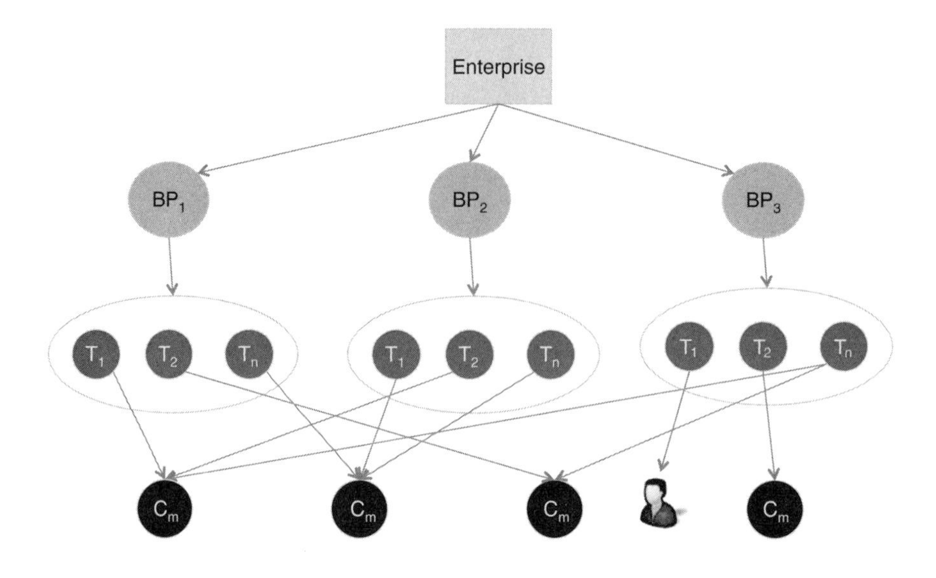

Figure 2.7 Business process orientation and technology

domain-specific tasks. Figure 2.7 illustrates this by connecting many tasks across multiple business processes to the same technology component.

We are going to think of these technology components in abstract terms consistent with a model-driven architectural pattern but with one difference: we are going to reduce such a model to metadata to eliminate programming altogether (Figure 2.8). For example, if the task is to send data to another

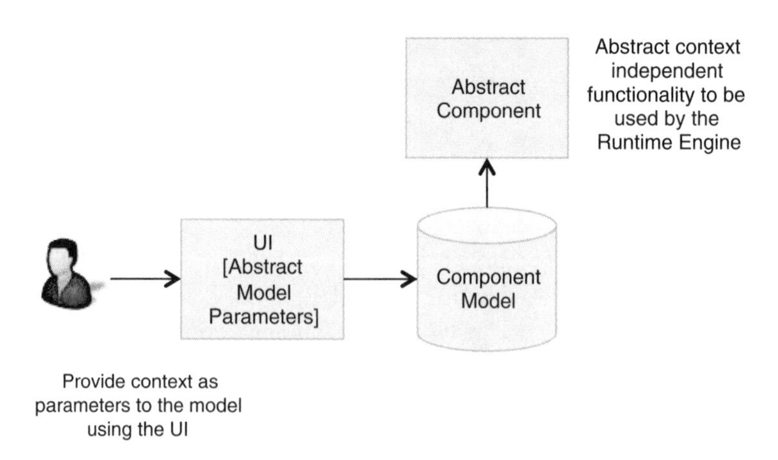

Figure 2.8 Anatomy of an abstract component

system through a data file, then we can think of an abstract technology component called "connector." The connector component should be able to facilitate connectivity between systems not just for this specific task but for any such system to system interface task.

Let's delve a bit deeper into the abstract connector component. All system to system interfaces move data from a source system to a target system using one of several transport protocols. The data from the source to the target might have to be transformed in some way to fit the target data structures. We can then create a comprehensive abstraction of the functionality required to implement a generic interface between two systems with a given protocol. Such an abstraction becomes the component's model. The abstraction is devoid of any context, only the abstract functionality required to move data from one source to another. This abstraction will contain no knowledge of the source or target or their data structures. Such an abstraction can then be exposed for use in specific development projects through a set of properties of the model, also referred to as metadata. To implement a specific interface, the user will provide details of the source, target, source file, target file, any transformation between source and target, and so on. Very important is to note that such context-specific details will all be in the form of data to the connector component's abstract model.

The purpose of the abstract connector component is to let the business analyst model a specific interface using the model of the abstract component. If the model is comprehensive, the implementation of an actual interface can be reduced to pure modeling. Such modeling involves providing as data, parameters that will be the specifics of the interface the analyst is interested in implementing. The actual interface is thus reduced to parameters to the model of the abstract component.

To truly enable a business achieve process orientation, we need a technology platform that has an extensive set of such abstract components. This set must also include components for knotty knowledge-based tasks, such as extraction of data from semi-structured and unstructured documents. The platform will have an orchestration component that will allow the designer to knit together business processes end to end using the appropriate collection of abstract technology components. With a meta model-driven architecture, a context is provided to each of these abstract components purely in the form of data using their respective abstract models. Such contextual business process models can be modified at any time. Note we are not just modifying the workflow like BPM software but the entire business process, including any/all of its tasks.

We believe such a BPA framework can provide the efficiency and agility that every enterprise needs to be competitive. The enterprise can now truly differentiate itself with its expertise.

2.5.2 RAGE AI™

RAGE AI™ (Rage) is a knowledge-based automation platform that has been designed to rapidly create flexible, process-centric software applications entirely through modeling. The objective behind the Rage platform is to create "dynamic" applications on a near real time basis purely through modeling and without programming. At the same time, it is extensible to let development teams add any functionality that cannot be modeled using the available components. The platform was founded on the belief that intelligent automation is the basis of systemic competitive advantage. The platform can enable enterprises to transition to a true process-centric paradigm without a lot of change management. *As processes are institutionalized in RAGE AI™, the RAGE AI™ machines can be the process glue.*

RAGE AI™ embodies the belief that business owners need to be able to modify business processes near real time to rapidly meet constantly changing business needs, and cost effectively, without worrying about technology nuances. The platform derives its process centric orientation from the process model of a business described in Figure 2.7. The RAGE AI™ platform currently provides a set of 20 abstract components, referred to as "engines," all using a meta model-driven architecture. The platform's runtime orchestration engine and all its abstract components work with metadata and do not translate any of the process models into code.

RAGE AI™ envisions that as users implement a specific application domain, they will discover common subprocesses and tasks that can all be isolated and very easily made reusable, even across domains within the firm.

As shown in Figure 2.7, the Rage methodology facilitates the decomposition of a business process into a set of tasks and enables the RAGE AI™ modeler to map each task to one of the 20 engines. Thus, implementation of the task is reduced to providing contextual parameters to the appropriate abstract component's model. The process orchestration engine enables the modeling of the orchestration again using the orchestration engine's model. This is figuratively shown in Figure 2.9.

Virtually, any change in the application can be easily handled by changing the appropriate models without worrying about previously generated code and deployment architecture. Thus, RAGE AI™ solutions are always "live,"

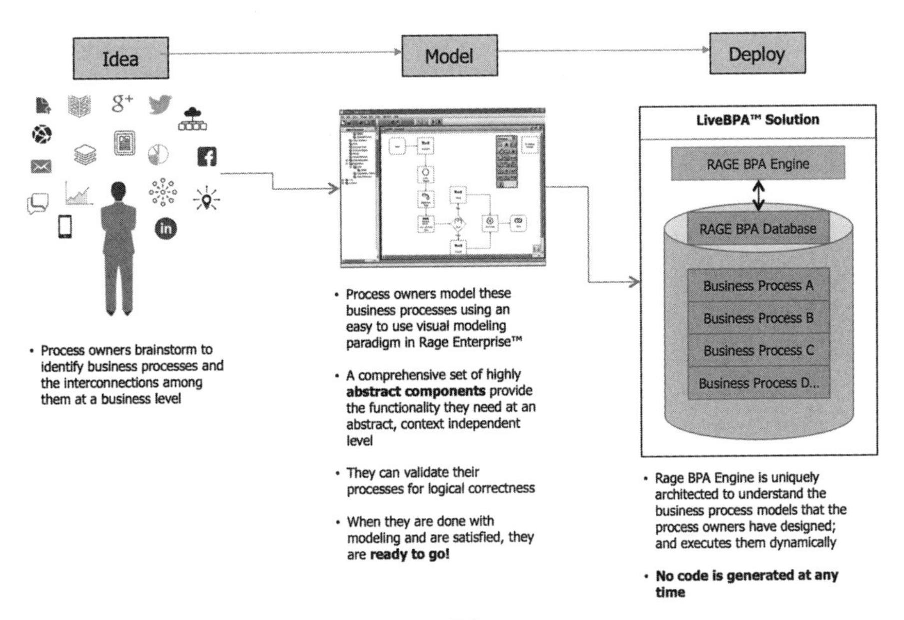

Figure 2.9 RAGE AI™ solution architecture

capable of being modified easily, rapidly and without any maintenance overhead. At the same time, Rage is an extensible platform and easily integrates with legacy application[s]. Application developers can add legacy code segments as custom tasks and extend the functionality of the framework.

2.5.3 RAGE Abstract Components

The RAGE AI™ platform comprises 20 engines at time of writing this book. Figure 2.10 outlines all the engines with a brief description.

Each of the 20 components conform to a consistent model-driven architecture, allowing business analysts to use them by providing parameters to the models of the respective engine. These models remain the way they are structured.

In contrast, most of the BPM software provides 1 or 2 of the 20 engines. Many provide accelerators that are pre-built applications for specific verticals. With RAGE AI™, the entire application can be easily implemented at any time through modeling using the depicted 20 engines.

Let's revisit the earlier loan processing example outlined in Figure 2.4. Using LiveBPA, the leverage picture will look substantially different (Figure 2.11).

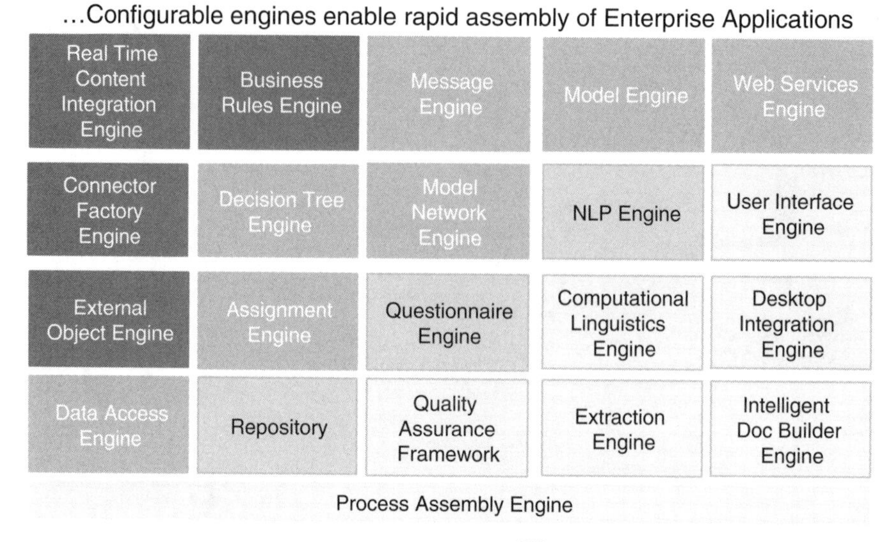

Figure 2.10 RAGE AI™ engines

All the tasks in the high-level flow can be fully supported by the meta model-driven RAGE AI™ platform compared to a very low level of coverage from the popular BPM platforms in the market. Figure 2.12 outlines the major types of functionality needed for each of the high-level tasks and the corresponding abstract components in LiveBPA that will be used to model these tasks:

As Figure 2.12 shows, to fully support the functionality, each task needs a lot more than a rules engine and a workflow orchestrator. Accelerators for specific vertical domains are of limited use because they will not provide the flexibility that a broader set of abstract components do.

2.5.4 RIM™ - An Actionable, Dynamic Methodology

The RAGE AI™ solution lifecycle for software development, RIM™, is dramatically shorter compared to a contemporary SDLC. RIM™ is an Agile

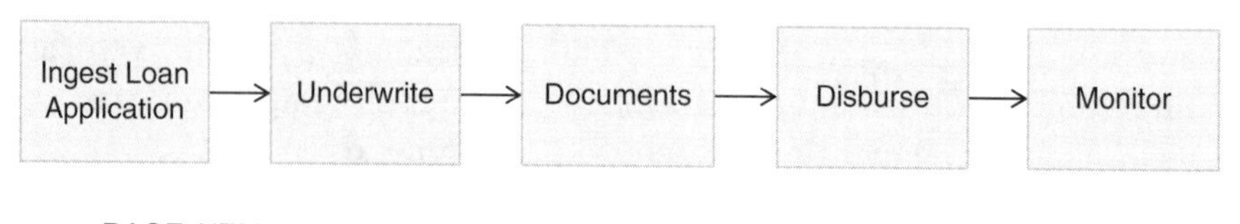

Figure 2.11 Machine intelligence role in E2E loan processing application

Task	Major Functionality	Illustrative RAGE AI™ Components
Ingest Loan Application	Online Application Form Case Assignment	UI, Rules, Process, Decision Tree
Underwrite	Access Credit Info Normalize Credit Info Spread Financial Info Determine Credit Limits Price Credit Communicate Decision	UI, Rules, Process, Connector for Credit Info Access, Connectors for Integration with Internal Systems, Extraction for semi- or unstructured documents, Computational Network for Credit Pricing Models, Decision Tree
Documents	Generate Appropriate Docs Create PDF Electronic Signatures	UI, Rules, Questionnaire, Document Generator
Disburse	Disburse Funds	Connector for Funds Gateway Integration
Monitor	Monitor the Riskiness of the Borrower on an Ongoing Basis	Connector for Integration with External Sources, Natural Language Processing, Deep Linguistics

Figure 2.12 High-level task to component map

methodology and can be compared to Scrum[1] at a high level. But because the software development process using the RAGE LiveBPA platform has a process-centric design focus and fundamentally does away with coding, there are significant differences. RIM enables development teams to implement orders of magnitude more functionality in the same time as Agile methodologies because implementation is reduced to modeling using abstract components and there is no translation to code.

In RIM, requirements and implementation are largely synonymous. Besides, "models" are the implementation. Just like XP and Scrum, there is no emphasis on detailed documentation. Once the models are complete, they are tested for functional correctness. Traditional syntactical testing is limited to the custom code being developed, if any.

The "Design" step in RIM is quite novel. The overall requirements are decomposed into a constituent set of business processes, called a "Domain." A functioning storyboard is rapidly developed using the LiveBPA™ platform. A functioning storyboard is similar to a prototype, with the key difference that the storyboard is a working version and can be thought of as a very early alpha version of the functionality.

[1] Scrum, Extreme Programming (XP) are popular Agile methodologies.

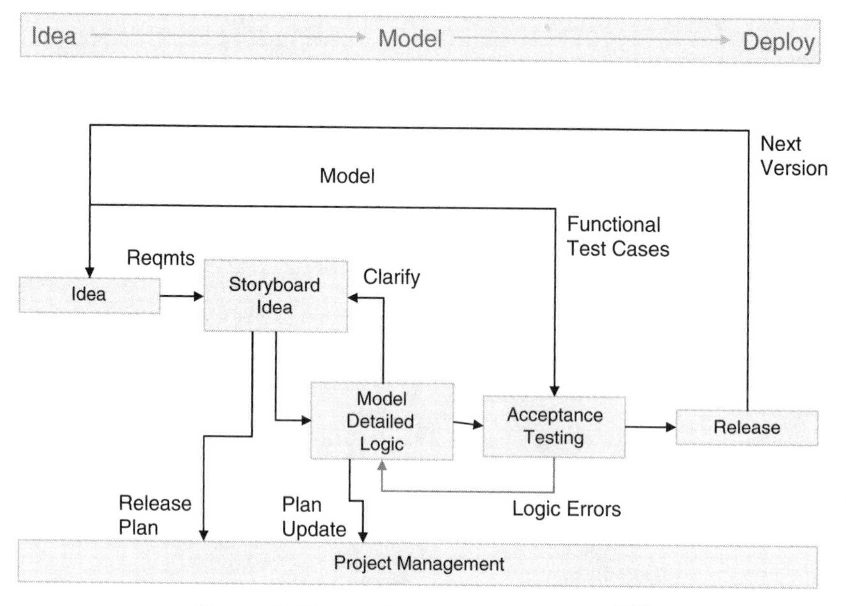

Figure 2.13 RAGE methodology: RIM™

Figure 2.13 introduces a set of new concepts that are explained below briefly.

Idea A high-level statement or diagram of the scope or idea.

Domain The scope or idea decomposed into high-level business processes that constitute the idea. This is a critical step because it forces the institutionalization of domain expertise.

Storyboard A functioning high-level prototype of the Domain. It is different from a conventional prototype. A functioning storyboard eventually evolves into the application. A storyboard can contain one or more user stories. The key in a user story is obviously that it should be minimally viable. Too many Agile implementations don't pay attention to this basic step. The critical need for an expert is also frequently overlooked.

Models The business processes that are part of the Domain and implemented by the implementation team using RAGE AI™. Most application functionality can be modeled using the extensive set of components in the RAGE AI™ platform. Depending on the size and scale of the effort, the modeling work can be distributed among the business process teams for implementation.

Functional tests Applications of sets of interrelated business process models that are stored as data. There is no need for conventional testing. What needs to be tested is functional accuracy of the models.

Project management RIM updating of project management steps. In addition to the overall project plan, daily project updates ensure that any emerging risks are identified as soon as they present and suitable corrective steps are taken by the project team. The initial project plan is finalized once the storyboard is ready. If necessary, the project plan is updated as the storyboard models acquire more detail and lucidity.

After a release, the cycle repeats all over again.

RIM is superior to generic Agile methodologies only because of the underlying RAGE AI™ platform. The platform allows the methodology to reduce/eliminate the non-value-added stages of a conventional SDLC like programming, syntactical testing.

2.5.5 Real Time Software Development

The previous discussion outlined the power of RAGE AI™ with an extensive set of abstract components that allowed a firm to rapidly and flexibly implement mission critical software applications purely through modeling. However, the RAGE AI™ solution lifecycle does not stop there. It extends its meta model-driven architectural paradigm quite dramatically all the way to requirements modeling.

As Figure 2.14 illustrates, RAGE AI™ provides a comprehensive modeling environment for visual modeling of requirements. The requirements model relies on the business process decomposition structure in Figure 2.7. The modeler can model requirements using high-level palette of abstract business tasks and provide the relevant task level information as parameters to the abstract tasks.

At the time of gathering requirements, the user will know the underlying information needs of the various business tasks though a formal canonical data model may not yet be available. The modeler should be asked to list the information needs of each task along with attributes, like cardinality. An abstract data modeling component then intelligently converts such an information model across all the business processes into a draft canonical model. Such an auto-generated data model can be refined by a data architect if necessary.

Once the required models have the required data references, another intelligent component in RAGE AI™ converts them automatically into implementation models. The conversion is accomplished by automatically mapping

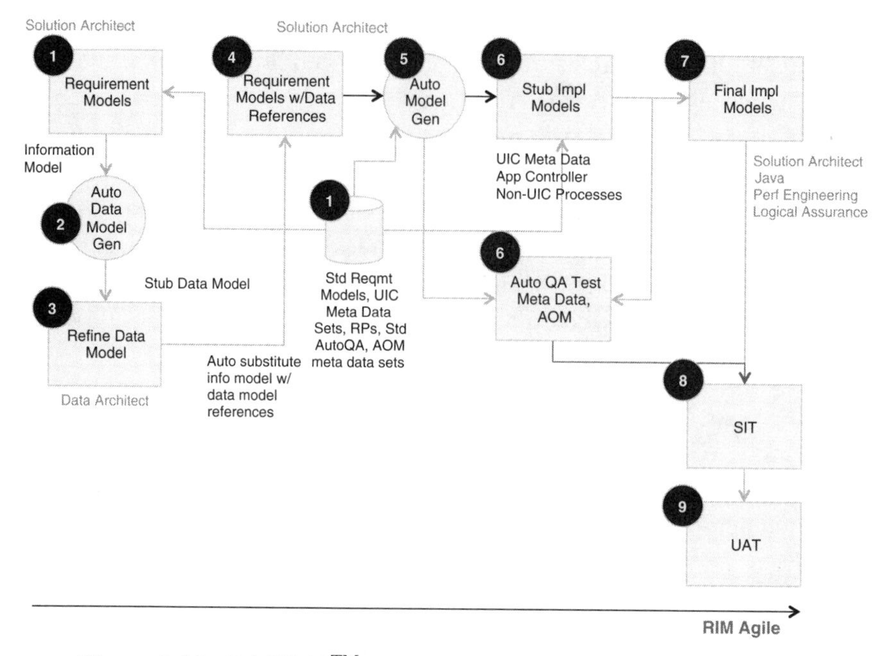

Figure 2.14 RAGE AI^{TM} - Real time software development lifecycle

each of the tasks in the requirement models to one or more components in RAGE AI^{TM}. The conversion component will also auto-generate the parameter values required for the implementation models with appropriate context specific parameter values. The table in Figure 2.15 briefly defines each of the numbered steps in the RAGE AI^{TM} lifecycle.

The RAGE AI^{TM} platform, as the figure shows, can enable near real time software development to eliminate and substantially reduce programming. The platform achieves perpetual flexibility through its meta model-driven architecture.

2.6 SUMMARY

We have been long debating methods that ensure that enterprises become efficient and flexible. Process orientation, as opposed to a functional organization, was widely touted as the best re-engineering solution, but most organizations have not succeeded in becoming process oriented. All the while, as this change is still being contemplated, technology has become the ever more pervasive enabler of business needs.

Step	Title	Resources/Tool	Description
1	Requirement Models	RAGE, Solution Architect	Model requirements using RIM and list the information model
2	Auto Gen Data Model	RAGE	Automated data generator component will convert the information model to a relational data model
3	Refine Data Model	RAGE, Data/Solution Architect	Any changes needed to the data model; as much as possible change the info model specs; should go away in the long run
4	Requirement Model w/Data Flows	RAGE	Automated data model component to substitute info model references to appropriate data model references
5	Auto Gen Models	RAGE	Auto generator to generate QA test meta data, initial version of the implementation models including UI, overall app controller, AOM data
6	Initial Impl Models	RAGE	Auto generated impl models, Appl Object Model (AOM), User Interface Component (UIC)
7	Initial AutoQA Tests	RAGE	Auto generated QA test meta data
8	SIT	Solution Architect, Project Manager	Systems Integrated Testing
9	UAT	Solution Architect, Project Manager	User Acceptance Testiing
10	Reusable repository of UIC meta data, auto QA meta data, processes, etc	RAGE, MBT	

Figure 2.15 RAGE AI™ real time software development lifecycle steps

However, despite the increasing criticality of technology, its historic inability to be responsive to business needs is the Achilles heel in the quest of businesses to become highly efficient and agile. Agile methodologies do not dramatically make the software development cycle faster or more flexible. Even in incrementing efficiency, Agile methodologies need well-trained experts to be on hand to guide the changes, and most organizations do not have the personnel trained in using Agile. Also Agile ends up lengthening the time to market and does not solve the fundamental issue of flexibility. As a result most organizations have not fundamentally altered the forward engineering paradigm employed in contemporary software development.

We present an entirely new platform called RAGE AI™. This is a revolutionary enterprise application development platform that can disrupt the traditional software development lifecycle by orders of magnitude. With its extensive set of abstract meta model-driven components, RAGE AI™ reduces the software development lifecycle, all the way from requirements elicitation to implementation and to a visual modeling exercises. The model in fact is the application in RAGE AI™. We believe that RAGE AI™ can truly make enterprises become efficient and agile.

REFERENCES

Ang, S., and Straub, D. (1998). Production and transaction economies and IS outsourcing: A study of the U.S. banking industry. *MIS Quarterly* 22(4): 535–552.

Armistead, C. (1996). Principles of business process management. *Managing Service Quality* 6(6): 48–52.

Armistead, C., Pritchard, J.-P., and Machin, S. (1999). Strategic business process management for organisational effectiveness. *Long Range Planning* 32(1): 96–106.

Armistead, C., and Machin, S. (1997). Implications of business process management for operations management. *International Journal of Operations and Production Management* 1(17): 886–898.

Attaran, M. (2003). Information technology and business-process redesign. *Business Process Management Journal* 9(4): 440–458.

Balzarova, M. A., Bamber, C. J., McCambridge, S., Sharp, J. M. (2004). Key success factors in implementation of process-based management: A UK housing association experience. *Business Process Management Journal* 10(4): 387–399.

Bloch, M., Boskovic, D., and Weinberg, A. (2009). How innovators are changing IT offshoring. *McKinsey Insights*, October.

Buciuman-Coman, V., and Sahlean, A. G. (2005). The dynamically stable enterprise: engineered for change, BPMGroup. In *Search of BPM Excellence: Straight from the Thought Leaders*. Tampa: Meghan-Kiffer.

Burlton, R. T. (2001). *Business Process Management: Profiting from Process*. Indianapolis: SAMS.

Byrne, J. A. (1993). The horizontal corporation. *Business Week*: 76–81.

Castro, J., Kolp, M., Mylopoulos, J.: Towards requirements-driven information systems engineering: The Tropos project. *Information Systems* 27(6): 365–389.

Chung, L., Nixon, B., Yu, E., Mylopoulos, J. (2000). *Non-Functional Requirements in Software Engineering*. Dordrecht: Kluwer.

ComputerWeekly (2015). What to consider when IT outsourcing contracts come up for renewal. October 3.

Dardenne, A., van Lamsweerde, A., Fickas, S. (1993). Goal-directed requirements acquisition. *Science of Computer Programming* 20: 3–50.

Davenport, T. H. (1993). *Process Innovation: Reengineering Work through Information Technology.* Boston: Harvard Business School Press.

Davenport, Thomas E. (2005). The Coming Commoditization of Processes. *Harvard Business Review*, June.

Davis, B. (2004). Finding lessons of outsourcing in 4 historical tales. *Wall Street Journal*, March 20, A-1, A-8.

DeToro, I., and McCabe, T. (1997). How to stay flexible and elude fads. *Quality Progress* 30(3): 55–60.

DeVellis, R. F. (1991). *Scale Development: Theory and Applications*. Newbury Park: Sage Publications.

Economist, The. (2003). The new geography of the IT industry. July 17.

Economist, The. (2003). Lost in translation. November 27.

ET Bureau (2015). Wipro to reduce headcount with investments in automation, artificial intelligence and digital technology. April 28.

Finders, K. (2015). How technology is changing BPO. *ComputerWeekly* 2015. http://www.computerweekly.com/feature/How-technology-is-changing-BPO

Galbraith, J. R. (1995). *Designing Organizations, An Executive Briefing on Strategy, Structure, and Process*. San Francisco: Jossey-Bass.

Genpact (2015). Genpact introduces Lean Digital^{SM} – A unique approach that delivers Digital's full potential and helps reclaim over $400 billion of impact, New York: September 15. http://www.genpact.com/docs/default-source/pr/genpact-introduces-lean-digital-a-unique-approach-that-delivers-digitals-full-potential-and-helps-reclaim-over-400-billion-of-impact

Gustavo, A., Dadam, P., and Rosemann, M. (eds.). Business process management. In *Proceedings of the 5th International Conference, BPM 2007*. Brisbane, Australia, September 24–28.

Hammer, M. (1996). *Beyond Reengineering: How the Process-Centered Organization Is Changing Our Lives*. New York: HarperBusiness.

Hammer, M. (2007). The process audit. *Harvard Business Review* April: 1–14.

Hammer, M., and Champy, J. (1993). *Reengineering the Corporation: A Manifesto for Business Revolution*. New York: HarperBusiness; revised updated ed., HarperCollins, 2004.

Harmon, P. (2003). *Business Process Change: A Manager's Guide to Improving, Redesigning and Automating Processes*. San Francisco: Morgan Kaufmann.

Harmon, P. (2004). Evaluating an organization's business process maturity. *Business Process Trends* 2(3): 1–11.

Harrington, H. J. (1991). *Business Process Improvement: The Breakthrough Strategy for Total Quality, Productivity and Competitiveness*. New York: McGraw-Hill.

Hill, J. B., Dunie R., and Chin K. (2015). Critical capabilities for BPM-platform-based case management frameworks. Gartner Research, August 12.

Hung, R. Y. (2006). Business process management as competitive advantage: A review and empirical study. *Total Quality Management* 17(1): 21–40.

Indulska, M., Recker, J., Rosemann, M., and Green, P. (2009). Process modeling: Current issues and future challenges. *International Conference on Advanced Information Systems Engineering*. Berlin: Springer, 501–514.

Jochem, R., Geers, D., and Heinze, P. (2011). Maturity measurement of knowledge-intensive business processes. *TQM Journal* 23(4): 377–387.

Kaka, N. (2009). Strengthening India's offshoring industry. *McKinsey Quarterly*, August.

Kohlbacher, M. (2009). The effects of process orientation on customer satisfaction, product quality and time-based performance. *29th Annual International Conference of the Strategic Management Society*, 1–7.

Kohlbacher, M. (2010). The effects of process orientation: A literature review. *Business Process Management Journal* 16(1): 135–152. Twenty-Third European Conference on Information Systems (ECIS), Münster, Germany.

Kohlbacher, M., and Gruenwald, S. (2011). Process orientation: Conceptualization and measurement. *Business Process Management Journal* 17(2): 267–283.

Kohlbacher, M., and Reijers, H. A. (2013). The effects of process-oriented organizational design on firm performance. *Business Process Management Journal* 19(2): 245–262.

Kotter, J. P., and Heskett, J. L. (2003). *Corporate Culture and Performance*. New York: Free Press.

Lacity, M., and Willcocks, L. (1998). An empirical investigation of information technology sourcing practices: Lessons from experience. *MIS Quarterly* 22(3): 363–408.

Lee, S. M., Olson, D. L., Trimi, S., Rosacker, K. M. (2005). An integrated method to evaluate business process alternatives. *Business Process Management Journal* 11(2): 198–212.

Lee, J., Lee, D., and Kang, S. (2009). vPMM: A value based process maturity model. In *Computer and Information Science* 2009, SCI 208. Berlin: Springer, 193–202.

Leymann, F. and Altenhuber, W. (1994): Managing business processes as an information resource. *IBM Systems Journal* 33(2): 326–348.

Marr, B., and Schiuma, G. (2003). Business performance measurement – Past, present and future. *Management Decision* 41(8): 680–687.

McCormack, K. P., and Johnson, W. C. (2001). *Business Process Orientation: Gaining the e-Business Competitive Advantage*. Boca Raton, FL: CRC Press.

McCormack, K. P., Johnson, W. C., and Walker, W. T. (2003). *Supply Chain Networks and Business Process Orientation*. Boca Raton, FL: CRC Press.

McKinsey (2009). Strengthening India's offshoring industry. August.

Moxon, R. (1975). The motivation for investment in offshore plans: The case of the US electronics industry. *Journal of International Business Studies* 6(3): 51–66.

Nandakumar, I., and Prasad, A. (2013). IT margins headed inexorably downwards, may fall below 20%. *Analysts*, The Economic Times, February 19. http://articles.economictimes.indiatimes.com/2013-02-19/news/37179764_1_margins-infosys-and-wipro-sid-pai

Object Management Group (OMG). (2008). *Business Process Maturity Model (BPMM)*, Ver.1. Needham, MA: OMG.

Pritchard, J.-P., and Armistead, C. (1999). Business process management – Lessons from European business. *Business Process Management Journal* 5(1): 10–35.

Reingold, J. (2004). A brief (recent) history of offshoring. *Fast Company*, April 1.

Rosemann, M. and de Bruin, T. (2005). Towards a business process management maturity model. In *ECIS 2005 Proceedings. Paper 37.* http://eprints.qut.edu.au/25194/1/25194_rosemann_2006001488.pdf.

Rosemann, M., de Bruin, T. and Power, B. (2006). BPM Maturity. In *Business Process Management: Practical Guidelines for Successful Implementations,* J. Jeston and J. Nelis, eds. Amsterdam: Elsevier.

Rummler, G. A. and Brache, A. P. (1990). *Improving Performance: How to Manage the White Space in the Organization Chart.* San Francisco: Jossey-Bass.

Rummler, G. A., Ramias, A. J., and Rummler, R. A. (2006). Potential pitfalls on the road to a process managed organization (PMO). *Business Process Trends* 2(3): 1–13.

Sawhney, M. (2002). What lies ahead: Rethinking the global corporation. Digital Frontier Conference, September 6.

Smith, H., and Fingar, P. (2003). *Business Process Management: The Third Wave.* Tampa: Meghan-Kiffer.

Soon, V. (2015). Cognizant's investments in disruptive technologies paying off: Francisco D'Souza. *LiveMint*, May 5. http://www.livemint.com/Companies/hbmArfACTaTPPlVtAZbRrJ/Francisco-DSouza–Investments-in-disruptive-technologies-p.html.

Srinivasan, V. (2009). The BPO market is not sustainable in the long run. *Knowledge@Wharton*, December 17. http://knowledge.wharton.upenn.edu/article/rage-frameworks-venkat-srinivasanthe-bpo-market-is-not-sustainable-in-the-long-run/.

Tarhan, A., Turetken, O., and Reijers, H. A. (2015). Do mature business processes lead to improved performance? – A review of literature for empirical evidence. Discussion Paper 178. *In 23rd European Conference on Information Systems (ECIS)*, Munster, Germany.

Tenner, A. R., and DeToro, I. J. (2000). *Process Redesign: The Implementation Guide for Managers*. Englewood Cliffs, NJ: Prentice Hall.

Tonchia, S., and Tramontano, A. (2004). *Process Management for the Extended Enterprise: Organisational and ICT Networks*. Berlin: Springer.

Towers, S., Lyneham-Brown, D., Schurter, T., and McGregor, M. (2005). 8 Omega, BPMGroup. In *Search of BPM Excellence: Straight from the Thought Leaders.* Tampa: Meghan-Kiffer Press.

van der Aalst, W. M. P., and van Hee, K. M. (2002). *Workflow Management: Models, Methods, and Systems.* Cambridge: MIT Press.

Wang, T. G. (2002). Transaction attributes and software outsourcing success: An empirical investigation of transaction cost theory. *Information Systems Journal* 12: 153–181.

Willaert, P., and Bergh, J. Van Den (2007). The process-oriented organisation: A holistic view developing a framework for business process orientation maturity. *BPM 2007, LNCS 4714*. Berlin: Springer, 1–15.

Zairi, M. (1997). Business process management: A boundaryless approach to modern competitiveness. *Business Process Management Journal* 3(1): 64–80. http://www.prnewswire.com/news-releases/genpact-introduces-lean-digitalsm—a-unique-approach-that-delivers-digitals-full-potential-and-helps-reclaim-over-400-billion-of-impact-300143843.html

CHAPTER 3

INSIGHT AND INTELLIGENCE

3.1 INTRODUCTION

From efficiency and effectiveness, we transition to insight and intelligence. The focus in this chapter is on the quantum of unstructured data at our disposal and the need for a contextually relevant framework to uncover intelligence from such information. We begin by highlighting the information overload prevalent today. We examine why timely access and analysis of information is important. And we discuss why the value of differential information is the key competitive factor in most industries.

We then proceed to discuss the popular methods for creating machine intelligence and knowledge representation. We point out why "black box" methods are unlikely to be widely accepted. We present a framework that can enable machine learning and knowledge representation in a contextually relevant, traceable manner. We end the chapter with lists of several real world applications that are in use today by thousands of people.

For the reader with a technical background, further technical details on the various methods for machine learning beyond the overview provided in this chapter can be found in the references at the end of the chapter.

The Intelligent Enterprise in the Era of Big Data, First Edition. Venkat Srinivasan.

3.2 THE EXCITEMENT AROUND BIG DATA

What do we mean by "big data"? Why is there so much hype about big data? What opportunities does big data afford enterprises?

Digital Intelligence Today (2013) reports the following facts about the digital world:

- 90% of all the data in the world has been generated over the last two years.
- Information consumption in the United States is in the order of 3.6 zettabytes (3.6 million million gigabytes).
- Twenty-eight percent of office workers' time is spent dealing with emails.
- The human brain has a theoretical memory storage capacity of 2.5 petabytes (petabyte = million gigabyte).
- The maximum number of pieces of information a human brain can handle is 7 (Miller's law).

Similarly, in their annual digital universe study, IDC finds (Gantz and Reinsel, 2012):

- At the midpoint of a longitudinal study starting with data collected in 2005 and extending to 2020, our analysis shows a continuously expanding, increasingly complex, and ever more interesting digital universe.
- From 2005 to 2020, the digital universe will grow by a factor of 300, from 130 exabytes to 40,000 exabytes, or 40 trillion gigabytes (more than 5,200 gigabytes for every man, woman, and child in 2020). From now until 2020, the digital universe will nearly double every two years.
- While the portion of the digital universe holding potential analytic value is growing, only a tiny fraction, less than half a percent, has been explored according to IDC. IDC estimates that by 2020, as much as 33% of the digital universe will contain information that might be valuable if analyzed, compared with 25% today.

Unstructured text makes up a disproportionately large share of the new information we have access to. By some estimates, 90% of all data on the Internet is unstructured text.

What does all this tell us? Enterprises not only have an enormous opportunity to gain insights from all this unstructured data, but this will become a competitive necessity to analyze all unstructured data for all possible insights.

Big data has existed in each phase of the computing evolution. Our perspective on big data has been shaped by the computing infrastructure available to us. In the early days of computing, even small amount of data (by today's standards) seemed very big. There are many businesses that have always had significant transaction data, such as financial institutions like American Express or Citigroup, or information companies like Dun & Bradstreet. While these companies can claim to have dealt with big data all the time, there has been a quantum, nonlinear leap in the amount and type of data available to us today because of the Internet and social media.

Today, big data is popularly defined to comprise four dimensions: Volume, Variety, Velocity, and Veracity. A decade ago, enterprises planned for data storage in terabytes and that was considered a big leap from gigabytes. Now, we routinely talk of petabytes and the speed with which we are producing petabytes of information is shrinking fast (*Fortune*, 2012). Big data goes well beyond traditional structured data to include a variety of unstructured data – text, audio, images, and so on. A large portion of the explosive growth in data is unstructured. There are two aspects to the Velocity dimension – the huge growth in the movement of data around the world that is requiring much bigger and massively parallel computing infrastructure, and the increasingly real time nature of the information and its processing/consumption. Unlike traditional structured data where the sources of data were known and accurate, a significant amount of big data today comes from unknown and unreliable sources and presents nontrivial accuracy and credibility challenges.

Beyond the massive expansion in the amount of data and the computing power available today, the excitement around big data revolves around two interrelated dimensions in my view:

1. The ability of computers to analyze "full" data sets, and where necessary in real time, instead of limiting ourselves to samples. We now have the computing power to analyze complete information instead of limiting ourselves to samples and making assumptions about population characteristics and how well the sample represents the population. Frankly, most people use parametric statistics blindly without regard to the suitability of the assumptions they are making about the sample and the underlying population. The ability to analyze the entire population frees us from all those arbitrary and sometimes unrealistic assumptions!
2. Growing ability of computers to analyze the underlying structure of unstructured text! Interestingly, unstructured text does not really mean

that there is no structure to the text. Any text using a formal language is, by definition, structured according to the rules of that language! So by unstructured text, we really don't mean that the text has no structure, only that the structure is not readable or understandable by the computer employing techniques routinely used before.

These two developments have dramatically shifted our ability to gain insight and intelligence from all the data unlike anything before. In the next section of this chapter, we provide a brief overview of the challenges related to information overload and asymmetry that motivate us to use AI to harness the intelligence in all data. We provide a brief overview of statistical machine learning methods both for structured data and for natural language. These methods, with one exception, are largely black boxes and are void of any contextual reasoning or theory. We introduce a powerful framework, using computational linguistics and decision trees, that is transparent in its reasoning, can evaluate data in context, and can reflect prior knowledge or theory.

3.3 INFORMATION OVERLOAD, ASYMMETRY, AND DECISION MAKING

3.3.1 Information Overload

Information overload simply refers to "too much information." Eppler and Mengis (2004) provide a comprehensive review of different types of overloads and ongoing research. Among the types of information overload already identified by the academic community are cognitive overload (Vollmann, 1991), sensory overload (Libowski, 1975), communication overload (Meier, 1963), knowledge overload (Hunt and Newman, 1997), and information fatigue syndrome (Wurman, 2001). Other studies have attempted to situate and validate types of information overload within professional services, ranging from auditing (Simnet, 1996) to strategizing (Sparrow, 1999), business consulting (Hansen and Haas, 2001), management meetings (Grise and Gallupe, 1999/2000), and supermarket shopping (Jacoby et al., 1974; Friedmann, 1977).

So far, such studies have defined information overload in different contextually relevant ways and measured their impact on decision accuracy. For example, researchers in marketing have compared the volume of information with the information processing capacity of an individual. Information overload occurs when the supply exceeds the capacity. Another related way

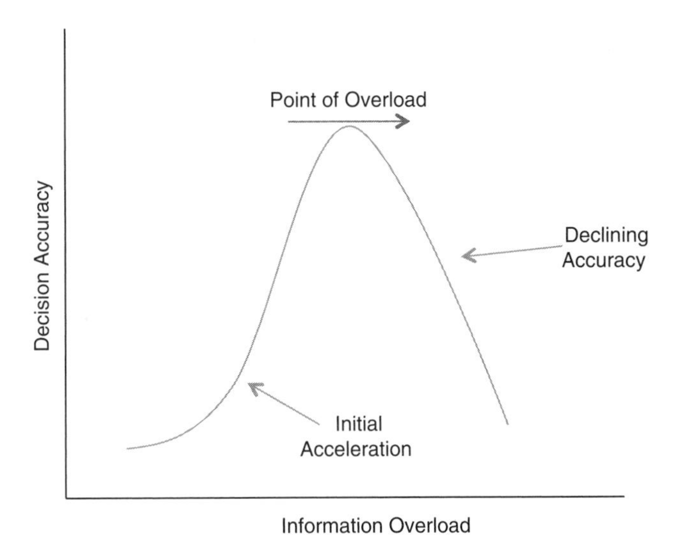

Figure 3.1 Information overload and decision accuracy

compares the quantity of information that an individual can integrate into his or her decision-making process within a defined time period with the amount of information that needs to be processed. In this regard, "time" is considered the most important component, rather than the characteristics of the information. All these studies have established that information over has a positive impact on the accuracy and speed of human decision making up to a certain limit and, beyond that, a negative impact. Schick et al. (1990) argue that information overload can confuse the individual, affect his or her ability to set priorities, and makes prior information harder to recall.

Figure 3.1 shows the likely relationship between information and decision accuracy, all other things being equal.

As information accumulates, we can expect the rate of decision accuracy to increase and rise up to a point where the value of incremental information begins to decline, as shown in Figure 3.1. Eventually, excessive information will have a negative impact on decision accuracy. Schroder et al. (1967) were first to suggest an inverted U-curve for depicting the relationship between decision accuracy and information overload.

Several authors (Schneider, 1987; Simpson and Prusak, 1995; Sparrow, 1998) have since pointed out that the specific characteristics of information – such as the level of uncertainty associated with information, the level of ambiguity, novelty, complexity, and intensity – have differential impacts on the information processing capacity of an individual and therefore information overload.

Moreover, it is important to recognize that with information overload the well–documented sequential position effect (Coleman, 2006) could become quite pronounced. The sequential position effect is the tendency of a person to recall the first and last items much better than the middle items. With information overload, the number of middle items will obviously be very large and likely especially pertinent to the decision making. Given such primacy and recency effects, the information that arrives in the middle of the decision maker's processing window may well be ignored.

Thus, as can be intuitively understood, too much information impacts decision-making accuracy. It is therefore important that some way be found not to miss critical information and to ensure that decisions are based on all the available information.

3.3.2 Information Asymmetry

While information overload refers to all the information a human mind has to process and its impact on decisions made by humans, information asymmetry refers to the difference in information provided or available to two or more individuals. Information asymmetry is at the heart of all transactions and interactions between individuals and corporations.

Information asymmetry has been the focus of considerable academic research in corporate finance with respect to financial markets and investors. Regulatory bodies like the Securities and Exchange Commission have enforced regulations whose purposes are to ensure that there is transparency and uniformity in any information shared by public entities. While such symmetric release of information is necessary to protect investors, it is also information asymmetry that is the basis of all commerce. When one individual has a different set of information from another, it creates the basis for a difference in perceived value creating an opportunity for trade.

Information asymmetry can be of many types. Figures 3.2, 3.3, and 3.4 depict three types of asymmetry. Figure 3.2 shows asymmetry due to differential access to information. This can happen for a number of reasons, including information overload. If you have too much information to process and cannot get to the important piece of information that somebody else has, there is information asymmetry between you and the other person.

Figure 3.3 shows another type of access-driven information asymmetry. In Figure 3.3, two individuals A and B process different non-overlapping information on the same topic. Such an asymmetrical information state can occur when one individual cannot access critical information for whatever reason.

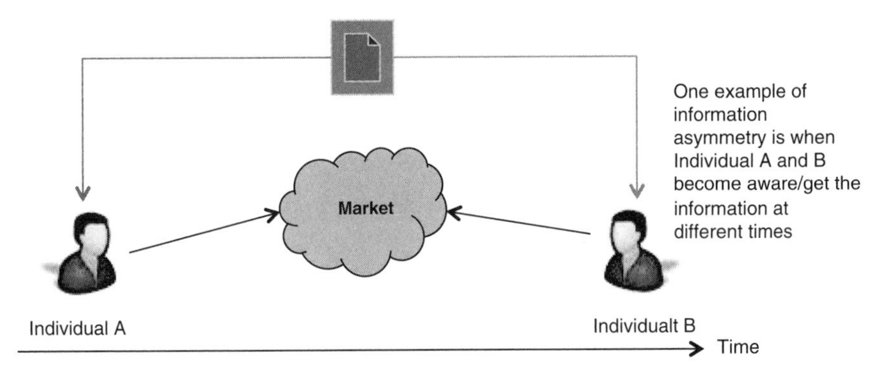

Figure 3.2 Access-driven information asymmetry

Asymmetry will be evident in A and B potentially reaching a different conclusion about the topic. If the topic is a company, A and B might potentially have different views of the value of the company and that can impact trade in financial or asset markets.

Figure 3.4 shows a type of information asymmetry based on an individual's interpretation of the information at his or her disposal. Interpretation reflects, of course, the individual's competence and understanding of the

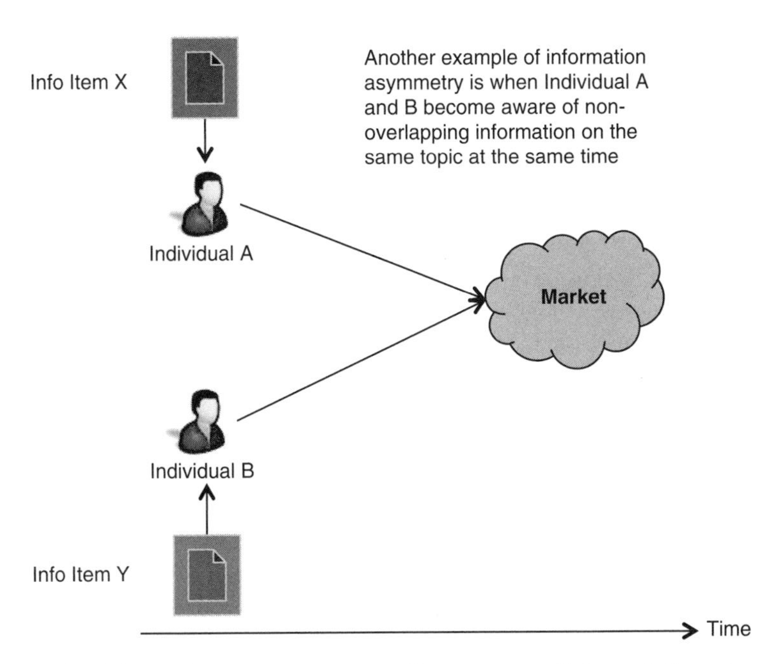

Figure 3.3 Information asymmetry caused by access to different information at the same time

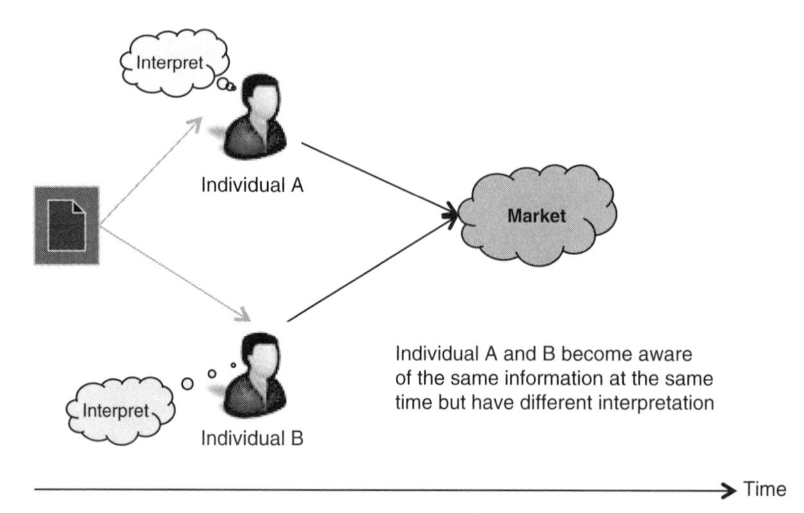

Figure 3.4 Information asymmetry caused by differences in interpretation of information

topic. Individuals A and B might have levels of job experience that can influence the way they interpret the information. Differences in interpretation can also occur when the interpretation calls for forecasting business trends. Forecasting involves a lot of knowledge extraneous to the supplied information and thus presents another type of asymmetry. Forecasts are usually based on assumptions about future states, with differences in assumptions impacting the information being interpreted and other information that individuals A and B might need to access in engaging in their commercial transactions, which could be buying and selling, directly or through a market mechanism such as the stock market.

Of course, information asymmetry can be intentional. Individual A may feed misinformation to individual B so that B will act on insufficient or inaccurate information (Figure 3.5). Much has been written about the imbalance of information distributed in certain industries. Consumers too have have very little access to objective information.

Generally, this type of information asymmetry is what regulatory bodies are committed to reduce or remove. Many companies, especially start-ups, are attempting to remove or reduce such asymmetry by providing consumers with the same amount of information that sellers have. For example, FindTheBest offers widely ranging information on consumer products, including cars, smartphones, and colleges. In the popular literature there are additionally many other examples of information asymmetry that has been shattered by technology (Stefano, 2006).

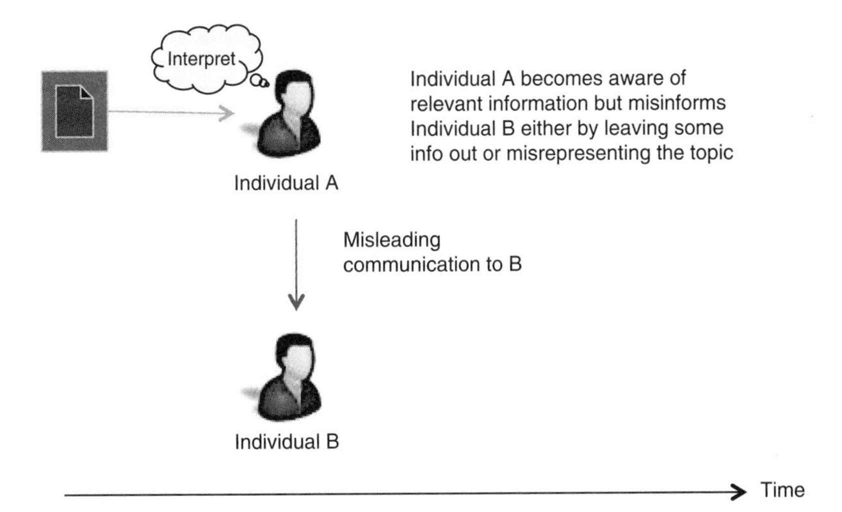

Figure 3.5 Information asymmetry caused by intentional misinformation

It is important to understand that the types of asymmetry illustrated in Figures 3.2, 3.3, and 3.4 are those commonly necessary for commerce. In fact, all information asymmetry is not bad; some forms of asymmetry are essential.

To be competitive, enterprises are constantly trying to find ways of differentiating themselves and their product value from that of their competitors. An important enabler of such competitiveness is the insight that enterprises can glean from the data available to them. The more asymmetry they can create in their favor without intentionally misleading other market participants, the more competitive advantage they will have created for themselves.

Overall, the huge explosion in data has exacerbated the information overload challenge and has created opportunities for enterprises to raise the information asymmetries between themselves and their competitors, and necessary in order to create or maintain their competitive advantage. The field of artificial intelligence has stepped in to provide solutions both to create competitive advantages and to overcome any competitive disadvantage due to information asymmetry.

3.4 ARTIFICIAL INTELLIGENCE TO THE RESCUE

Can artificial intelligence (AI) help handle the exponential growth in information, turn overload into an advantage by uncovering critical bits of information without human intervention and leveraging knowledge and experience of humans? There is considerable evidence that it can!

AI as a field of study refers to machines or systems that can simulate knowledge-based work performed by humans. Researchers in the field define AI as the study and design of intelligent agents whereby the agents can act on data and reach decisions just like a human would in a rational environment (e.g., see Poole et al., 1998; Nilsson, 1998; Russell and Norvig, 2003; McCarthy, 2007). It is an interdisciplinary field drawing on statistics, computer science, mathematics, psychology, linguistics, and neuroscience.

3.4.1 A Taxonomy of AI Problem Types and Methods

The core objective of AI is simply to make machines so much more intelligent that they approach human intelligence. We can parse this core objective into multiple dimensions (as in Figure 3.6) that include reasoning, knowledge representation, automated learning, natural language processing, perception, and the ability to move and manipulate objects (Russell and Norvig, 2003; Lugar and Stubblefield, 2004; Poole et al., 1998).

Knowledge acquisition and representation is at the heart of machine intelligence. We humans over the course of our lifetimes accumulate enormous knowledge, both factual and judgmental. A key aspect of machine intelligence is to acquire and/or embed such knowledge in machines. The basic method used to promote machine acquisition of knowledge is through deductive and inductive reasonings.

Machine learning refers to computational algorithms that discover and improve knowledge through an algorithmic analysis of patterns in information. An important aspect of machine learning is the ability of the machines to update such intelligence with or without expert human assistance.

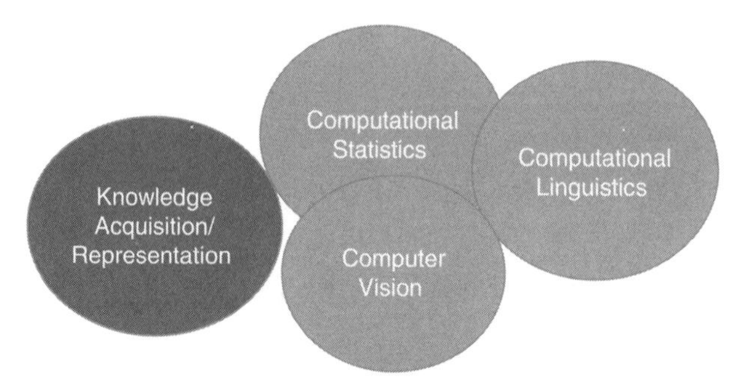

Figure 3.6 Dimensions of artificial intelligence

Natural language processing (NLP) addresses the need for machines to be able to read and understand the languages that humans speak or write in. Such needs can then be used in situations ranging from simple indexing of documents to interpretation and generation of documents just like humans do these things. Some people are beginning to refer to this as natural language understanding (NLU). NLP or NLU is machine learning focused on natural language. NLU is the focus of computational linguistics.

Machine perception or computer vision is the ability to ingest data from sensors of physical objects, such as cameras, microphones, drones, home equipment, and medical devices, to deduce physical dimensions and properties implied by the data. Closely related is the ability to handle physical activities like motion and navigation that call for machines having the capabilities of understanding localization, mapping, and motion planning (Russell and Norvig, 2003; Poole et al., 1998). Computer vision is also a form of machine learning focused on motion and perception.

Different business problems require machines to be capable of one or more of these dimensions. As machines become more intelligent with respect to these, they will become increasingly capable of making intelligent decisions in different application domains. Of course, this is a vast and complex endeavor with many dimensions, subject areas, and complexities. *In this book we are going to limit the discussion to the first four dimensions mentioned above; we will exclude computer vision though it is also a vast and very current area of interest.*

3.4.2 AI Solution Outcomes

As we mentioned above, AI applications focus on developing the intelligence and insight that machines can infer from data with or without human assistance. The various dimensions of artificial or machine intelligence can be translated to the type of solutions that we want intelligent machines to support. Figure 3.7 presents a taxonomy of such artificial intelligence solution outcomes. We can broadly categorize AI solution outcomes into five types: clustering, extraction, classification, prediction, and interpretation. It is also useful to broadly categorize these solution outcomes in terms of what they involve. Parametric statistical approaches attempt to model data based on pre-defined, explicitly programmed models and algorithms like logistic regression, multiple discriminant analysis based on strict assumptions about the data in the broader population. Machine learning methods attempt to let the data suggest the model and typically process the entire population of data instead of

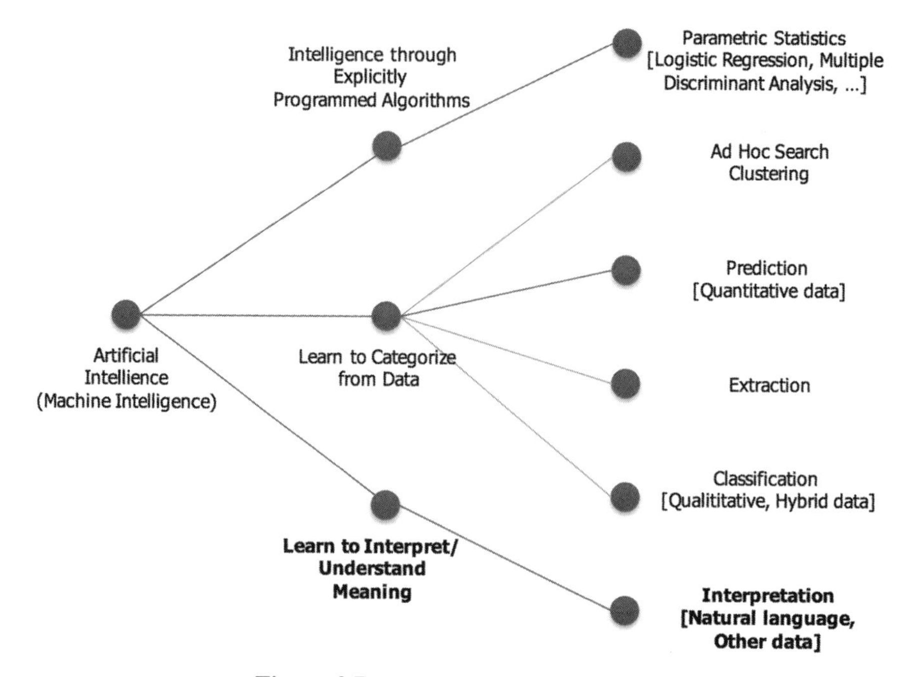

Figure 3.7 Taxonomy of AI solutions

drawing a sample. Finally, as it relates to natural language text, a key solution outcome is the need to interpret and understand the meaning of the text.

Clustering refers to the process of finding clusters of objects that are similar to each other and distinguished from data that are dissimilar. Business domains such as legal discovery, customer complaints, and ad hoc search can benefit from effective clustering of input data. Clustering is mostly performed to gain insight about the data, and this is obviously an iterative process. It is closely related to classification.

In Figure 3.8, four distinct sets of objects are arranged to illustrate the clean separation that occurs through clustering. Two clusters – A and B – are illustrated, and the line between them is the separating distance between them. In practice, real world data sets are much messier in that the objects are all intermingled and not as cleanly separable as illustrated in Figure 3.8. Clustering of real world data sets requires iterative parameter setting and fine-tuning.

Classification refers to the objective of classifying a document or entity into one of *N* pre-defined categories. Examples include predicting whether a company is heading toward bankrupcy or is healthy, classifying a document into a topic that it is about, classifying a customer complaint into a service category, and so on. A very large number of AI applications, do involve

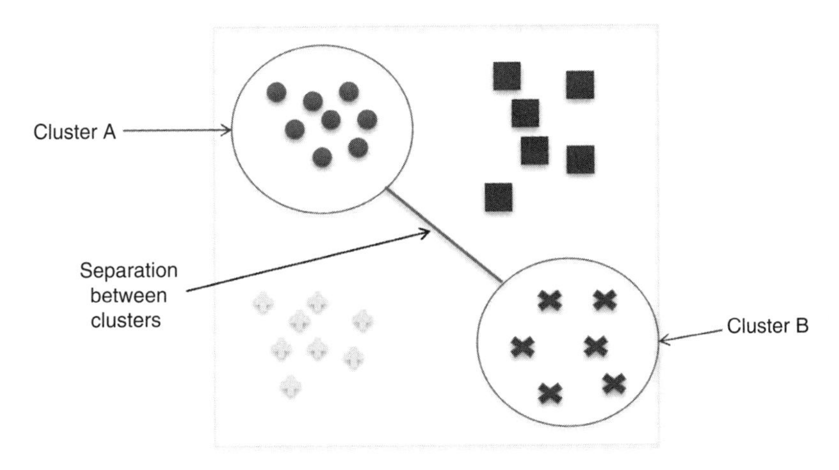

Figure 3.8 Clustering outcome type

classification. Classification is a natural extension to clustering. The major difference is that in classification, we know the categories ahead of time and in clustering we discover the categories from the data. Classifiers are developed based on a sample and then used to predict the category of a new item.

Figure 3.9 illustrates classification based on two variables – 1 and 2 – and two categories – A and B. In this simple case, the data self-assemble neatly into two distinct spaces. As mentioned in the case of clustering, real world data sets are much more complex and require a fair amount of data analysis and fine-tuning of the models to achieve meaningful classification accuracy.

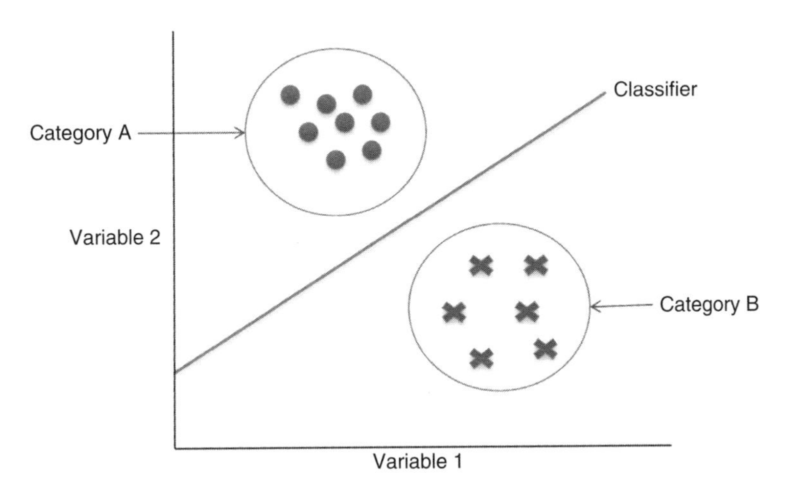

Figure 3.9 Classification outcome type

For most classification problems, there are four aspects. The first is a set of "independent" variables that can be categorical [0, 1,...] or continuous. This variable set is assumed to be causally related to the outcome, referred to as the "dependent" variable. In a classification outcome, the dependent variable is always categorical. An example of an dependent variable is bankruptcy prediction where we are interested in whether or not a business entity will be bankrupt. The second is the sample data, also referred to as the training dataset. A training data set will contain values for both the dependent and independent variables for a number of observations representative of the population for which the classifier is to be built. A third aspect is a validation data set, consisting observations on which the classifier can be validated. Last, the fourth aspect, is the cost of misclassification. Misclassification costs need not be the same for each outcome category. For example, the cost of classifying a healthy company as bankrupt may not be the same as the cost of classifying a bankrupt company as healthy, if you are a lender to the company. Some analysts also associate a prior probability with each outcome in the population. The classification model is called the "classifier."

Extraction refers to the need to extract specific data from documents. This need may manifest, for example, where tables of data in an unstructured or semi-structured document must be accessed and made actionable. A common example is financial data to be extracted from financial statements in a document such as an audit report. Since there are no standards for both the semantic and physical reporting of such data, extraction cannot use any location-specific pre-defined logic. Figure 3.8 shows a financial statement along with an excerpt from the footnotes. Since there is no standardized way of describing financial statement items, each company follows its own discretion in creating its financial statements – its chart of accounts, the format of the financial statement, and so on. The extraction problem here is to accurately extract the financial statement – labels, properties like statement as of dates, currency, unit, and the item values. Once extracted, these values and items have to be normalized at several levels.

Extraction as shown in Figure 3.10, might also involve the need to identify and extract breakups that might be listed somewhere else in the PDF document. As Figure 3.10 illustrates, the breakup for the item, Property and Equipment, is in a footnote to the financial statement.

Similarly, a very difficult business problem is the need to reconcile invoices with underlying legal contracts on which they are based. Figure 3.11 shows a legal contract and a screen from a RAGE AI™ application that includes some

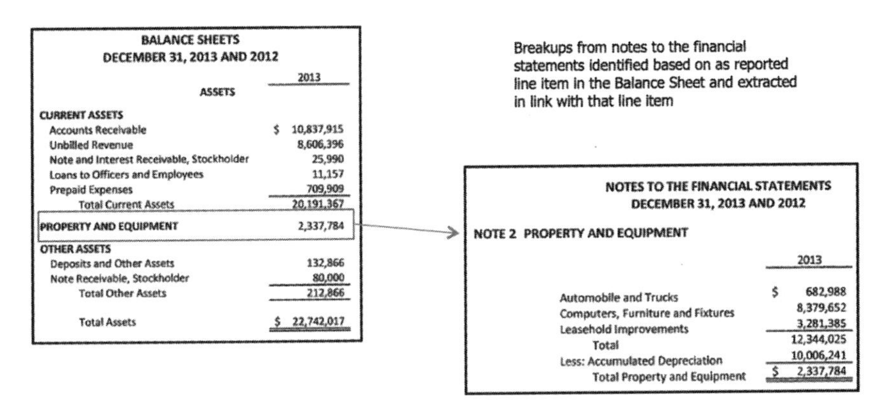

BALANCE SHEETS
DECEMBER 31, 2013 AND 2012

ASSETS	2013
CURRENT ASSETS	
Accounts Receivable	$ 10,837,915
Unbilled Revenue	8,606,396
Note and Interest Receivable, Stockholder	25,990
Loans to Officers and Employees	11,157
Prepaid Expenses	709,909
Total Current Assets	20,191,367
PROPERTY AND EQUIPMENT	2,337,784
OTHER ASSETS	
Deposits and Other Assets	132,866
Note Receivable, Stockholder	80,000
Total Other Assets	212,866
Total Assets	$ 22,742,017

NOTES TO THE FINANCIAL STATEMENTS
DECEMBER 31, 2013 AND 2012

NOTE 2 PROPERTY AND EQUIPMENT

	2013
Automobile and Trucks	$ 682,988
Computers, Furniture and Fixtures	8,379,652
Leasehold Improvements	3,281,385
Total	12,344,025
Less: Accumulated Depreciation	10,006,241
Total Property and Equipment	$ 2,337,784

Figure 3.10 Financial statement extraction from a PDF document

extracted values from the contract. In this case, a master services agreement is displayed and as a simple example, recruitment fee specified in a clause in the contract is extracted for comparison with an invoice.

Interpretation refers to the need to interpret unstructured content, within press releases, research reports, legal contracts, and so on, with respect to a certain purpose, which may be an investment, cost audit, sales prospecting, or any number of other things.

Figure 3.12 illustrates a news item on Walmart's attempt to launch a rival delivery system and unseat Amazon. The interpretation challenge in this

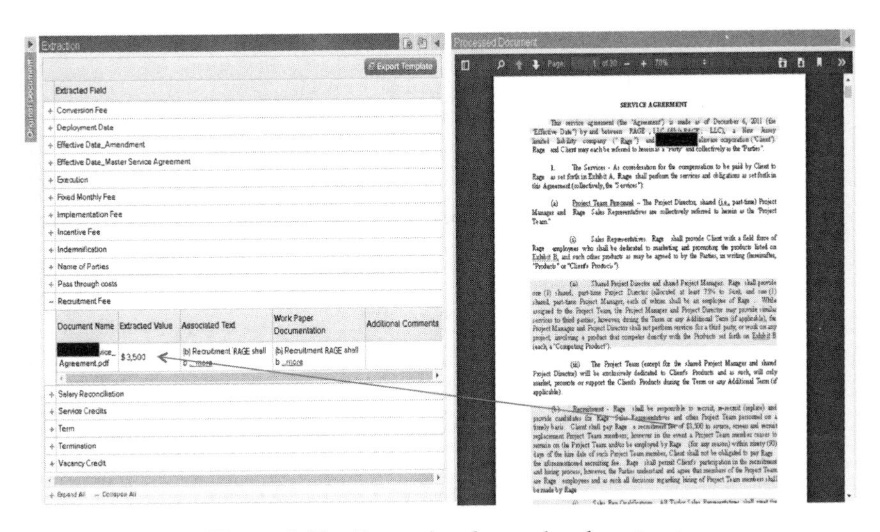

Figure 3.11 Extraction from a legal contract

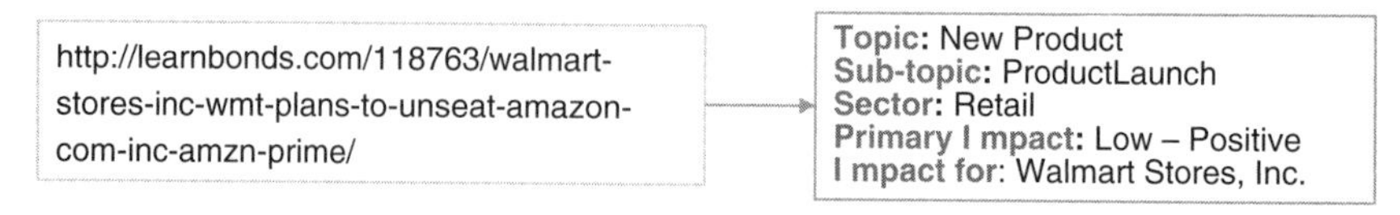

Figure 3.12 Interpretation outcome type

context is for the machine to understand this news item, that is, understand it just as well as you and I do. So the machine must first understand what the news item is about and what the news item is attempting to communicate. Then the machine needs to be able to understand how that could impact Walmart.

Note that in Figure 3.12 the desired outcome is that the news item be classified under the topic, "New Product," the sub-topic, "Product Launch," and the sector, "Retail." Further, note that the business problem requires the interpretation of the potential impact of this news item on Walmart. The impact in this case is judged by an analyst to be "low." To reach this judgment of the potential impact of this news item on Walmart, the entire news item needs to be read and interpreted in a global context surrounding Walmart.

Among the outcome types mentioned above, *interpretation* has received relatively little attention in the vast amount of literature that has appeared on machine learning, natural language processing, and sentiment analysis. We think this is largely because interpretation of natural language requires deep understanding of the language. Most work on machine learning so far has in fact been based on computational statistics and not on computational linguistics.

3.4.3 AI Solution Methods

A number of methods and tools have been developed to address the different problem types outlined in the previous section. These methods uncover insight and intelligence from the underlying data and embed that intelligence in machines. The methods range from parametric statistical methods like regression to nonparametric methods like support vector machines. Figure 3.13 maps common types of AI problems and the methods/tools that are used to solve them. We should also point out that for some such problems, more than one method might be used.

Interested readers are referred to Alpaydim (2010) for detailed explanations of the technical aspects of many of these methods.

Since in this book we are focusing on machine learning without preprogrammed algorithms based on assumptions about the data or the

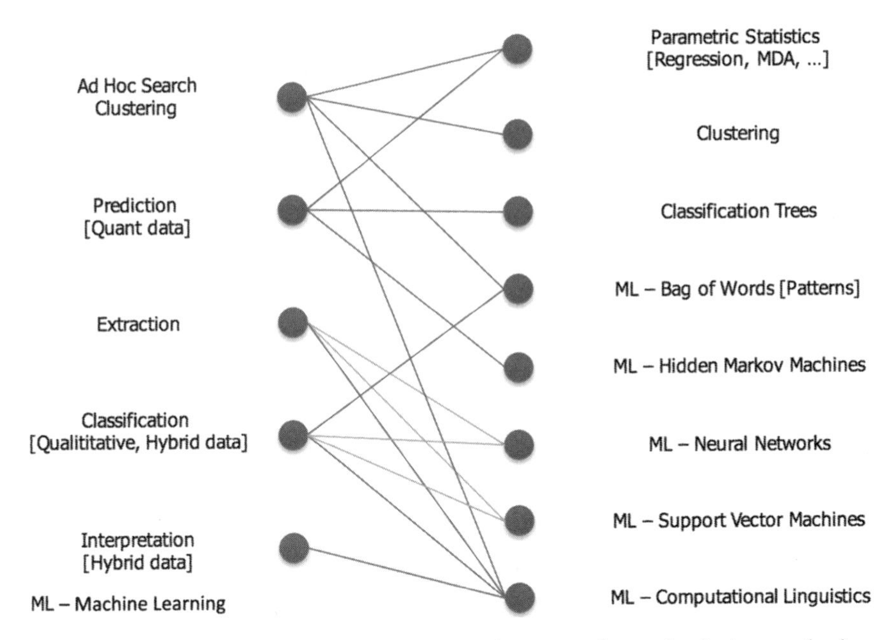

Figure 3.13 Map of AI problem types and commonly used solution methods

underlying population, for brevity, we will therefore group these methods into two categories – machine learning with computational statistics and machine learning for natural language with computational linguistics. As shown in Figure 3.13, all ML methods, except ML – computational linguistics – fall into the machine learning with computational statistics category.

Machine learning from data clearly suggests the availability of data. In classification problems, such data in some cases will also include data on the desired outcome, often called a "label." The data might be supplemented by human or expert knowledge. The knowledge type and how it is represented is a function of the AI problem type.

Machine learning tasks are typically classified into three broad categories as a function of the level of human or expert assistance available to the machine:

- *Unsupervised learning* No explicit "labels" are available and the learning algorithm has to find its own structure in the input data. Unsupervised learning is used as a means to an end.
- *Semi-supervised learning* Semi-supervised learning refers to some knowledge that is available to the machine learning algorithm where such

knowledge may not be complete. The learning algorithm has to combine such knowledge with other structure in the data to build its model.

- *Supervised learning* The machine learning algorithm has a data sample with input and desired output and has to build a model that will use the input data to predict or map to the desired output.
- *Reinforcement learning* A computer program interacts with a dynamic environment in which it must perform a certain goal (e.g., drive a vehicle), without a teacher explicitly telling it whether it has come close to its goal. Another example is learning to play a game by playing against an opponent. Recently, the Google Deepmind AlphaGo system claimed a historic victory against the Korean grandmaster, Lee Sidol, in Go, an exponentially more complex game than chess (Metz, 2016).

3.5 MACHINE LEARNING USING COMPUTATIONAL STATISTICS

It has been said that a computer program learns from experience; that is "... from experience E with respect to some class of tasks T and performance measure P if its performance at tasks in T, as measured by P, improves with experience E" (Mitchell, 2006, p. 1). Ever since Alan Turing (1950) proposed that we replace the question "Can machines think?" with the question "Can machines do what we (as thinking entities) can do?", the field has acknowledged a difference between operational terms and cognitive terms. We are, however, less convinced that there may be a real distinction between operational and cognitive terms.

Breiman (2001) suggests that there are two cultures in the use of statistical modeling to reach conclusions from data - the data model assuming that the data is generated by a given stochastic data model and the algorithmic model which treats the data as unknown.

Arguably, the data model, also referred to as parametric statistics, represents the earliest and simplest form of machine learning. Parametric statistical methods handled relatively simple interactions among data. Multivariate discriminant analysis was the earliest rigorous attempt to understand patterns in linear multivariate space that differentiated data belonging to different categories. Logistic regression was widely adopted as a way of overcoming the limitations of discriminant analysis. Srinivasan et al. (2009) use logistic regression to create a model to predict corporate bankruptcy. They demonstrate the efficacy of the logistic regression method by finding prediction accuracies of up to 90% for up to two years prior to bankruptcy.

Notwithstanding the success of these methods, parametric statistical methods suffer from significant limitations in modeling the full complexity of real world data and processes.

Modern machine learning (or the algorithmic model) has emerged as an interdisciplinary field from the study of patterns, computational learning theory in AI, mathematical optimization, parametric statistics, and computational linguistics. Machine learning explores the study and construction of algorithms that can be automatically deduced from data. Such algorithms learn by building a model from example data that generally include outcome data. They include supervised approaches such as artificial neural networks (Lek et al., 1996), cellular automata (Hogeweg, 1988), classification and regression trees (De'ath and Fabricius, 2000), fuzzy logic (Salski and Sperlbaum, 1991), genetic algorithms and programming (Stockwell and Noble, 1992), maximum entropy (Phillips et al., 2006), support vector machines (Drake et al., 2006), and wavelet analysis (Cho and Chon, 2006). In addition, there are unsupervised learning approaches that attempt to reveal patterns in data without any knowledge of output, including Hopfield neural networks (Hopfield, 1982) and self-organizing maps (Kohonen, 2001). These algorithmic methods can model complex, nonlinear relationships in data without the restrictive assumptions required by parametric approaches (Guisan and Zimmermann, 2000; Peterson and Vieglais, 2001; Olden and Jackson, 2002a; Elith et al., 2006).

We describe below briefly the more popular machine learning methods.

3.5.1 Decision Trees

Classification and regression trees (CARTs) have emerged in recent times as a powerful modeling approach to automatically develop classifiers in complex data sets especially where the predictive variables could be hypothesized to interact in a sequence, such as hierarchical. CART supports the belief that many real world outcomes are not caused by the simultaneous interaction of all the causal variables, a critical assumption in all of parametric statistics and in many of the modern machine learning methods described in this section.

CART has been widely used in a number of applied sciences, including medicine, computer science, and psychology (Ripley, 1996). De'ath and Fabricius (2000) recognize the advantages of CART when modeling nonlinear data containing independent variables that are suspected of interacting in a hierarchical fashion (De'ath and Fabricius, 2000). Olden et al. (2008) review a large number of CART applications in ecology. Frydman, Altman, and Kao (1985) applied CART to develop a predictive model of corporate bankruptcy.

CART is a binary recursive partitioning method that takes as input categorical and continuous variable data and outputs a decision tree as the classification model that classifies the observations into the desired outcome categories (Ogden et al., 2008). The model is built by splitting the observations into binary partitions based on a criterion using the independent variables. The observations, represented by a node in a decision tree, are split into two child nodes repetitively, including, if needed, the same variable again and again. Thus, each parent node can give rise to two child nodes, and in turn, each child node may itself be split, forming additional children. This splitting process results in a sequential and repeated partitioning of the data set allowing the resulting model to discover deeper classification patterns in the data.

Building a CART model involves instructions to the machine to start building a tree, stop the tree building process, and then allow a human pruning the tree to select the optimal tree. Tree building begins at the root node with the entire data set. All possible splitting conditions for each possible value for all independent variables are determined. The tree is then split by selecting the splitting condition that maximizes the "purity" of the resulting split, where purity is defined either as an information-based measure (entropy) or the Gini index for classification and sums of squares about group means for regression trees (De'ath and Fabricius, 2000). Tree building stops when the incremental gain from the next split of a node is not significant. Tree building can also stop with externally imposed constraints like a specific number of observations in each of the children nodes or the number of splits in the tree.

The selection of the final classifier (tree) could involve a pruning step where a human eliminates nodes of the classifier for reason of cost or complexity or for lack of causal justification. Classifier selection can also be automatically done through X-fold cross-validation (for a discussion on validation techniques and the optimal tree size, see Breiman et al., 1984; Bell, 1999; Hastie et al., 2001; Sutton, 2005). There are methods such as boosting (Shapire et al., 1998) and bagging (Breiman, 1996) that are aimed at generating more robust and generalizable trees automatically. Breiman (2000, 2001, 2004) demonstrated that using an ensemble of trees, where each tree is grown in randomly selected subsets of the data, can provide substantial gains in classification accuracy. Random forests, as they are called, have emerged as a powerful way to create decision trees that are more generalizable and robust.

Figure 3.14 illustrates a CART classifier for predicting corporate bankruptcy based on financial data. The data set and variables are described in Srinivasan (2009). The key variables that the classifier has selected are Adjusted Quick Ratio, Cash Flow from Operations to Sales, Debt to Total Assets, and Current Ratio.

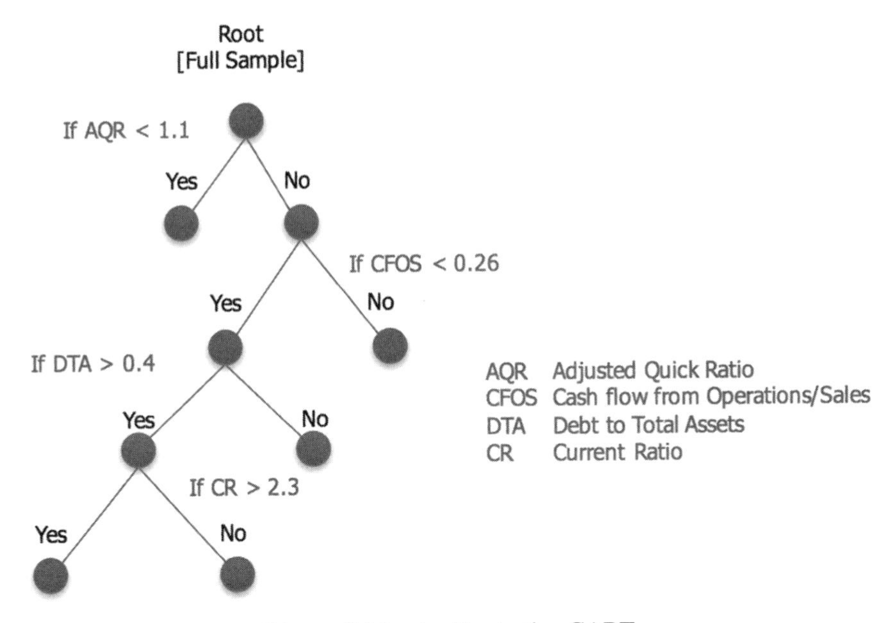

Figure 3.14 An illustrative CART

Three things should be pointed out regarding the CART classifier. First and foremost, it is traceable and transparent. We can see the variables and logic that make up the model and decide whether or not it fits our prior knowledge. It is not a black box. Second, the CART classifier is much simpler to interpret than other machine learning models and popular parametric statistical models like the multivariate logistic regression model. Third, the CART classifier is highly scalable, fast, and can handle a very large number of input variables without resulting in overfitted trees. Finally, the inherent "logic" in CART is readily apparent, and it makes financial sense.

3.5.2 Artificial Neural Networks (ANNs)

Artificial neural network (ANN) models, or, more generally, a multilayer perceptron, is a modeling approach reportedly inspired by the way the human brain functions (Alpaydim, 2010). The human brain is composed of a very large number of processing units (10^4). These processing units, called *neurons*, function by connecting, in turn, to approximately 10^4 other neurons through connections called *synapses*. In the human brain, these neurons operate in parallel.

In the human brain, the neurons communicate with each other through electrical signals that travel along the "axon" to the receiving neuron

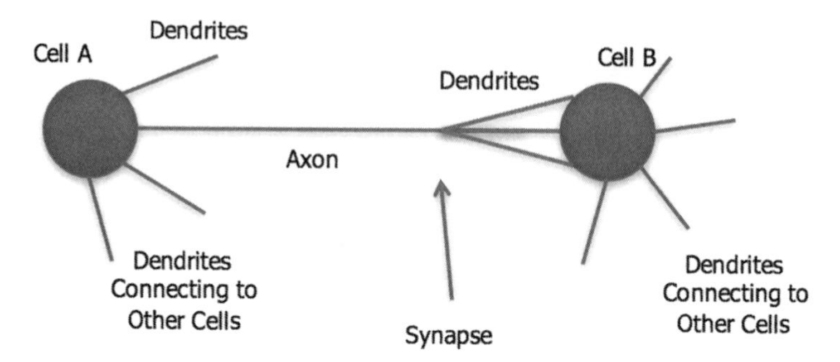

Figure 3.15 Neurons and synapses in the human brain

(Figure 3.15). The point at which the axon connects the receiving cell's dendrite has a wall called a synapse. Communication between the neurons happens through the release of different types of chemicals in appropriate quantities reflective of what is to be communicated.

The ANNs attempt to replicate this structure. While there are many supervised and unsupervised learning methods for ANNs (Bishop, 1995), we describe a very frequently used method: the one hidden-layer, supervised, feed forward neural network trained by the backpropagation algorithm.

As shown in Figure 3.16, the single hidden-layer feed forward network contains an input layer $[x_i]$, an intermediate non-linear transformation layer

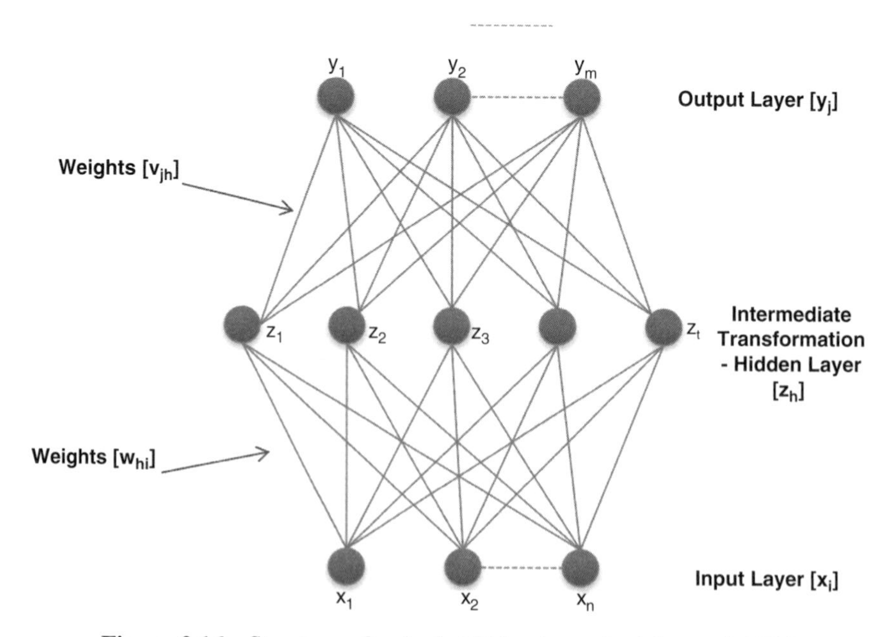

Figure 3.16 Structure of a single hidden-layer feed forward ANN

called the hidden layer [z_h], and an output layer [y_m], with each layer containing one or more neurons. There can be more than one hidden layer. With multiple hidden layers, the machine will attempt to model the successive hidden layer as a transformation of the previous hidden layer in a similar fashion. The number of neurons in the hidden layer can be selected arbitrarily or determined empirically by the modeler to minimize the trade-off between bias and variance (Geman et al., 1992). Just like other statistical models, the addition of neurons and layers will increase the model's precision in the training sample but will result in an over-fitted model that will yield a decrease in the recall of the model. A lower recall implies less generalizability.

Training the neural network typically involves an error backpropagation algorithm that searches for an optimal set of connection weights that produces an output signal with a small error relative to the observed output (i.e., minimizing the fitting criterion). For continuous output variables, the commonly used criterion is the least-squares error function, whereas for dichotomous output variables, the commonly used criterion is the cross-entropy error function, which is similar to log-likelihood (Bishop, 1995). The algorithm adjusts the connection weights in a backward fashion, layer by layer, in the direction of steepest descent, thus, minimizing the error function.

The training of the network is a recursive process where observations from the training data are entered into the network in turn, each time modifying the input-hidden and hidden-output connection weights. This procedure is repeated with the entire training dataset (i.e., each of the n observations) for a number of iterations or epochs until a stopping rule (e.g., error rate) is achieved. Prior to training the network, the independent variables should be converted to z-scores (0 to 1) in order to standardize the measurement scales of the inputs into the network.

As is the case with parametric methods, machine learning methods, including ANNs, use a training data set to estimate the models and then ascertain the model's fit using a test data set.

ANNs with a backpropagation-based feed forward approach have a few advantages over traditional parametric methods: (1) their nonparametric nature, in that they have no need for specific distributional assumptions of the independent variables, (2) ability to model nonlinear associations easily, and (3) the accommodation of variable interactions without a priori specification. Yet, ANNs suffer from some serious limitations in our view. The first and most important is that ANNs are largely a "black box." While there have been efforts to develop various ways to understand the contributions of the independent variables (e.g., see Olden and Jackson, 2002b), there is still no understandable explanation of the connection weights and layers. The adoption of such nontransparent methods especially for mission critical business applications will be challenging. Second, and a related issue, is the issue of causality, especially when several hidden layers are employed. Since variable interactions that the machine determines in the hidden layers are not predetermined, we run the risk of the model being very sample specific and identifying spurious statistical relationships. In CART, in contrast, we can observe the variable interactions and decide whether or not they make sense.

Kernel Machines

Kernel machines, also referred to as support vector machines (SVM), are supervised learning models with associated learning algorithms for classification and prediction. While the default SVM is a two-class algorithm, SVM can also be applied to multiclass problems. SVM attempts to create a classifier that divides the categories in the same data as widely as possible using the values for the causal variables that are provided. New examples are then mapped into that same space and predicted to belong to a category based on which side of the gap they fall on.

More formally, a support vector machine learns a separating hyperplane or a set of hyperplanes in high-dimensional space (the values for all the independent variables make up this high-dimensional space) that can be used for classification, regression, or other tasks. An effective hyperplane will maximize distance of the training data points of any class to the nearest category boundary. The training data set is generally specified with a fixed number of variables and data points; that is, it represents a finite-dimensional space. However, we know from experience that most real world data sets and phenomena are not so easily separable linearly in finite-dimensional space. Neither classifier A nor B separate the classes cleanly in Figure 3.17.

To address this issue, SVM transforms the original finite-dimensional space into an abstract higher dimensional space using a combination of the variable vectors using an algebraic vector multiplication method called "dot product." So, the independent variables will be combined to create new abstract variables. The idea is that this transformation will make separation a lot easier in the resulting higher dimensional space. The hyperplanes in the higher dimensional space are defined as the set of points whose dot product with a vector in that space is constant. The vectors defining the hyperplanes can be chosen as linear combinations with parameters α_i of variables, $x_{i.}$

One reasonable choice as the best hyperplane is the one that represents the largest separation, or margin, between the two classes. So we choose the

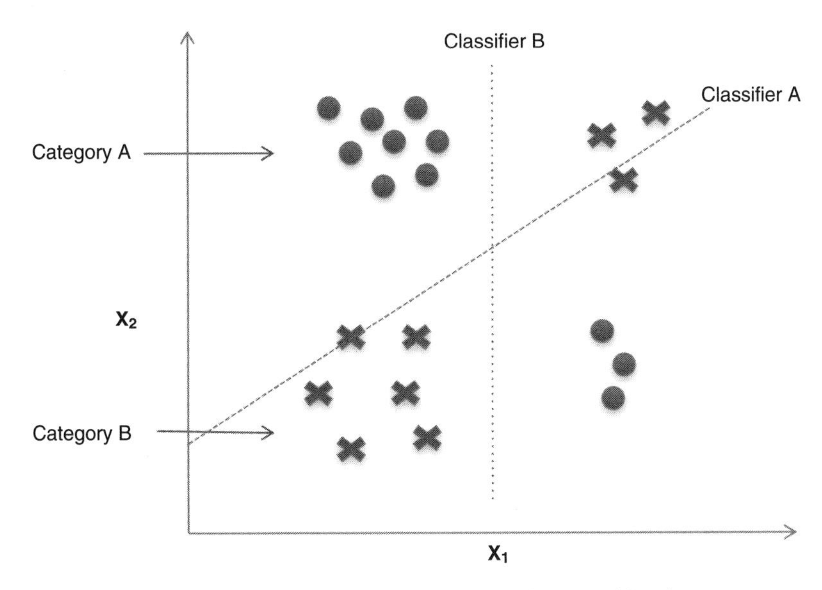

Figure 3.17 Geometric motivation for classification

hyperplane so that the distance from it to the nearest data point on each side is maximized.

Figure 3.18 illustrates a hyperplane that achieves the maximum separation for two classes A and B above.

If the training data are linearly separable, we can select two hyperplanes in a way that they separate the data and there are no points between them;

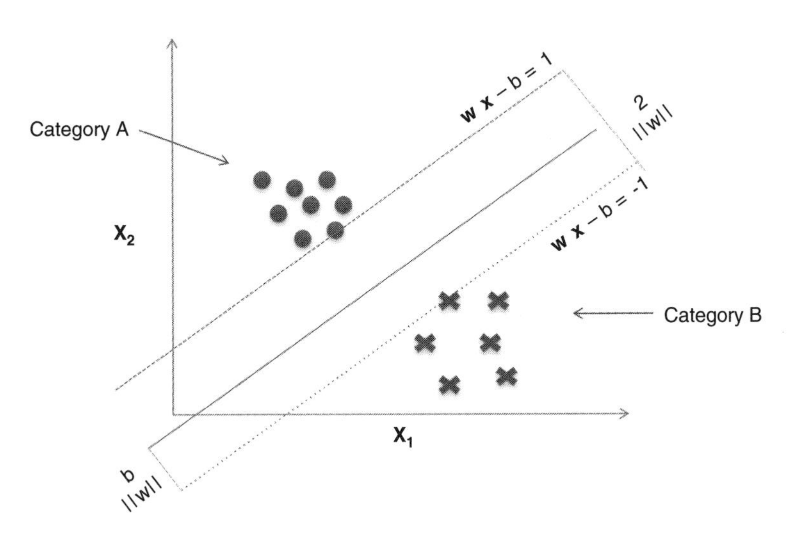

Figure 3.18 Illustrative SVM hyperplane separation

then we can try to maximize their distance. The data points that lie on the margins are called the support vectors. These hyperplanes can be described by the equations:

$$\mathbf{w} \cdot \mathbf{x} - b = 1$$

and

$$\mathbf{w} \cdot \mathbf{x} - b = -1$$

where $\cdot$ denotes the dot product and **w** the normal vector to the hyperplane. The parameter $\frac{b}{||w||}$ determines the offset of the hyperplane from the origin along the normal vector **w**.

The SVM classifier can be formulated as a convex quadratic programming problem that has many efficient algorithms. While the basic SVM can only be applied to two-class problems, this can be overcome by a variety of techniques. The most common of them is to reduce a multiclass problem into multiple two-class problems. Crammer and Singer (2001) proposed a multiclass SVM method that casts the multiclass classification problem into a single optimization problem, rather than decomposing it into multiple binary classification problems (Hsu and Lin, 2002).

Kernel machines are very similar to ANNs. They have the same strengths and weaknesses. The biggest weakness from our standpoint is the black box character of the model and the difficulties associated with interpreting the final model's parameters. This limits the adoption of the method to problems where understanding the model's components and their logical fit are not essential.

3.5.3 Deep Learning Architectures

Deep learning architectures are based on the assumption that observed data is generated by the interactions of many different factors at different levels. Deep learning algorithms have become quite popular especially in the field of computer vision, which includes such applications as autonomus driving and image and voice recognition. Convolutional deep neural networks are used in computer vision where their success is well documented (e.g., see Lecun, 1998, for an application to recognize handwritten character recognition). Bengio (2009) suggests that in order to learn the kind of complicated functions that can represent high-level abstractions, one needs multiple levels of nonlinear operations such as neural nets with many hidden layers.

The argument underlying deep learning is principally that in order to properly represent complex interactions, we need representational methods that mimic such multilevel, multilayer abstractions. Deep learning methods attempt to learn by building greedy models layer by layer attempting to disentangle complex abstractions. Deep learning methods are set up as unsupervised learning approaches, thus, obviating the need for large amounts of labeled data.

Generally, all deep learning methods suffer from the same weakness – lack of theory or explicit causality surrounding many of the methods. They are black boxes with most confirmations of their validity done empirically in the form of their results. They are fundamentally brute force methods that purportedly use replication of human brain function as the motivation for their architecture. In our opinion, while they are useful and can be highly effective in several classes of problems, their success depends on the availability of huge amounts of homogeneous data reflective of all possible cause-effect combinations including the associated context, which is rarely the case. In this sense, they do not truly replicate human brain functions.

The brain has a huge amount of stored knowledge and preferences that it draws on to act on the current stimuli. In the human brain, the neurons know what they are communicating to other neurons and why. The messaging among neurons is based on stored knowledge and not purely on some random combination of input variables. Humans have learned cause and effect over many years of experience and formal education. ANNs attempt to learn this purely from data. This will only work if the data is sufficiently exhaustive to reflect all the cause and effect relationships. Therefore, we believe that ANNs are likely to work where there is a relatively high level of homogeneity in the training data. For natural language processing, the missing piece in the deep learning methods based on computational statistics is the lack of contextual knowledge. In our view, the current deep learning architectures based on computational statistics do not truly replicate the way the human brain acquires and uses knowledge. They may appear to do so mechanically but are superficial in their replication of human brain functions.

As Marcus (2012) observes, deep learning the way it is being practiced so far is only part of the larger challenge of building intelligent machines. Such techniques lack ways of representing causal relationships, have no obvious ways of performing logical inferences, and they are also still a long way from integrating abstract knowledge, such as what objects are, what they are for, and how they are typically used. While IBM's Watson perhaps uses deep learning just as one element in an ensemble of techniques, all these criticisms apply.

3.6 MACHINE LEARNING WITH NATURAL LANGUAGE

Machine learning with natural language text involves two classes of problems – text classification and text interpretation. The bulk of the discussion in the literature with respect to natural language processing relates to text classification. There is an independent body of work related to sentiment analysis that is a narrow version of the interpretation problem. We discuss all of these in this section.

Many of the machine learning methods reviewed in the previous section can be used for text classification involving natural language. However, since text is different from numbers in representation, there are specific techniques to represent natural language text in a form that these methods can use to generate classifiers.

3.6.1 The "Bag-of-Words" Representation

The "bag-of-words" representation of natural language text has been the most popular way to reduce text to a form that statistical classifiers can use. In this method, text comprising a set of characters is transformed into a representation suitable for the learning algorithm and the classification task. Such a transformation relies on the well-known research finding that word stems can be a parsimonious representation construct. The word stem is derived from the occurrence form of a word by removing case and inflection information (Porter, 1980). For example "computes," "computing," and "computer" are all mapped to the same stem "comput" (Porter, 1980). Of course, this method assumes that the precise form of the word is less important in classification tasks.

The transformation reduces the text into a vector of unique words along with their occurrence frequency in the text. Each distinct word, w_i, is viewed as a Term [t] and the term frequency as *tf*. Term frequency is a way to weight terms differently to enable proper classification of a document to a class. Luhn (1957) came up with this definition of term weighting based on the simple assumption that the weight of a term in a document is proportional to its term frequency. For example, to find documents that can be classified properly in response to the query, "about artificial intelligence," it is not enough to simply find documents that have the three words, "about," "artificial," and "intelligence." We will still get a lot of irrelevant documents. To further filter the results, Luhn proposed that we count the number of times each term occurs in each document and sum them all together; the number of times a term occurs in a document is called its *term frequency*.

To measure the true information a word provides, and to minimize the impact of common words like "about" or "the" that do not provide any specific information, an inverse document frequency factor is used that reduces the weight of such commonly used words. Jones (1972) defined the inverse document frequency factor as an inverse function of the number of documents in which the word occurs. Another common technique to reduce the dimensionality of the word vectors is to remove words like "this," "it," "and," "or," and so on. These are referred to as "stop" words.

Each word, w_i, corresponds to a vector with $tf(w_i, d)$ that is the number of times word w_i occurs in the document d as its value. The inverse document frequency for a term, t, in a corpus of D documents is given by

$$idf\,(t, D) = \log \frac{N}{|\{d \in D : t \in d\}|}$$

where N is the total number of documents. The inverse document frequency for a term, t, is a measure of how much information that word provides. The *idf* diminishes the influence of a term that occurs very frequently and increases the weight of a term that occurs rarely. The inverse frequency of a word is low if it occurs in many documents and is highest if the word occurs in only one document.

Consider a document with 1000 words in which the word "intelligence" appears 50 times and a corpus of 1 million documents with the word "intelligence" appearing in 100 documents.

$$tf\,(\text{“intelligence”}) = 50/1000 = 0.05$$
$$idf(\text{“intelligence”}) = \log\,(1{,}000{,}000/100) = 4$$
$$tf - idf = 0.05 \times 4 = 0.20$$

There have been many attempts to explore more sophisticated techniques for representing text – based on higher order word statistics (Caroreso et al., 2001), NLP (Jacobs, 1992; Basili et al., 2000), "string kernels" (Lodhi et al., 2002), word clusters (Baker and McCallum, 1998), and graph-of-words (Rousseau et al., 2015), but the relatively simple "bag-of-words" has been the most popular.

Many of the computational statistical methods outlined in the previous section can generate classifiers based on the BOW representation. In fact, all of them rely on BOW.

3.6.2 Sentiment Analysis

Sentiment analysis of natural language, also referred to as opinion mining, is a popular and growing area of interest. Sentiment analysis involves the determination of "what other people think" or "what opinion has a person expressed." Sentiment analysis involves the identification of the overall sentiment expressed in an document regarding an item or a topic and/or opinion attributed to a specific stakeholder expressed within the document, typically as a quote from that stakeholder. Generally, this requires the identification of the area of the document where some evaluative text exists, attribution of that text to the item or topic or stakeholder, and interpreting the sentiment in that evaluative text. This type of analysis has been applied to movie reviews, product reviews and feedback, opinion blogs, book reviews, politics, and so on. In some cases, a source document makes the identification task much easier, such as book reviews on Amazon. We know they are book reviews and do not have to determine that from the text. However, free form text poses significantly more challenges.

Detection of the sentiment in a document involves the following aspects: First, we need to determine whether or not the document contains any sentiment. Second, assuming we have an opinionated document or text, we need to determine sentiment polarity – where does the sentiment lie on the continuum between positive and negative sentiment. Third, the document may contain sentiment on multiple topics, and we need to properly attribute the sentiment with its topic. Fourth, the document may contain sentiments expressed by multiple stakeholders or contain sentiments of the author on the topic and additional quotes reflecting the sentiments of other stakeholders. Finally, the sentiment may depend on knowledge or opinion outside the document.

Most sentiment analysis methods rely on term frequency (bag-of-words) or some variation of that, like term presence (Pang et al., 2002), position of a term within the textual unit (Pang et al., 2002), parts of speech (Mullen, 2004), and syntactical properties (Dave, 2003; Kudo and Matsumoto, 2004). Hatzivassiloglou and McKeown (1997) tried to group adjectives into two clusters such that maximum constraints are satisfied based on an hypothesis of how adjectives are separated. Wiebe (2000) analyzed adjectives for gradation and polarity. Statistical factors are used to predict the gradability of adjectives. Kim and Hovy (2004) relied on WordNet (Miller, 1995) to generate lists of words with positive and negative orientation based on seed lists. Such methods suffer from coherence weakness as synonyms using Wordnet do not account for the degree of similarity between the synonymous words and multiple senses. Godbole et al. (2007) proposed an alternate method to improve synonym set coherence in such discovery. Pang and Lee (2004) illustrated the use of minimal cuts in graphs to identify the subjective portions of a text span to categorize the overall sentiment contained in the text.

The focus of most sentiment analysis systems – including aspect-based analysis (Hu and Liu, 2004; Gamon, Aue, Corston-Oliver and Ringger, 2005; Carenini, Ng and Pauls, 2006; McDonald, Hannan, Neylon, Wells, and Reynar, 2007) – is highly limiting, with the general assumption that no prior knowledge of the domain being summarized is available. Further, none of these methods interpret the meaning of the text in a contextually relevant manner. As an example, the same discourse may carry a different sentiment for two different audiences.

We view the sentiment analysis problem as a subset of the text interpretation problem. The more holistic problem definition is one of how to understand the meaning of the text as humans do, whether it is for analyzing sentiment, or opinion, or for understanding the impact of facts and opinions in a document on a specific topic. The sentiment analysis methods and applications are narrowly focused on one aspect of understanding. By generally focusing only on the sentiment in the document being analyzed, these methods find only local (suboptimal) solutions. They are incapable of finding global sentiments incorporating prior knowledge and opinions outside of the document.

For machines to analyze natural language text, we need to embed in them a deep understanding of natural language and all its facets. They not only need to be able to process the text syntactically, they also need to be able to process it semantically like we humans do. Most online text processing including search engines, and other retrieval, aggregation, and processing applications are primarily based on algorithms applying the syntactical properties of the natural language text. These algorithms are very efficient and good at breaking up the text, checking spelling, parsing the texts, tracking the occurrence of words and word sequences, finding specific keywords, and other simple tasks. They mostly rely on word co-occurrence or n-grams, or variations thereof. These algorithms are very weak in terms of understanding the meaning of the text. In fact, their capabilities are severely limited in this regard.

We humans process text not just based on the written word. When we process text, our brain activates a lot of information that is not written. Many areas of our brain are activated when we read written text. We effortlessly recall our understanding of the word precisely in the context in which it is used wherever we can. In the process of understanding, the brain resolves any ambiguities associated with the usage of different words, any previous experiences we might have had related to that word and the concepts that it evokes in our brain. We understand the sentiment of the writer by relating to our sensory experiences and mapping of the words to them.

3.6.3 Knowledge Acquisition and Representation

A key aspect of natural language processing is "knowledge." Each of us continuously accumulates knowledge. For machines to approach human intelligence in processing natural language, there needs to be an effective way of acquiring, representing, storing, and adding knowledge to the machine. Knowledge acquisition can be unassisted, partially assisted, or provided entirely by experts. This has been discussed earlier in this chapter. Acquired knowledge has to be stored and represented inside the machine so it can be easily accessed.

Several knowledge representation schemes have been proposed over time. One popular representation scheme has been first-order logic (FOL) (Barwise, 1977). FOL is a deductive system consisting of axioms and rules of inferences that can include syntactic, semantic, and even pragmatic expressions. Syntax refers to the way groups of symbols are to be arranged in order to be considered "well formed," such as in a properly constructed sentence in the English language. Semantics refers to the meaning of such well-formed expressions. Pragmatics specifies how idiomatic and practical usage can be leveraged to improve semantics, such as the use of "xerox" to mean "copy." However, FOL is known to suffer from a level of rigidity in knowledge representation. FOL can be supplemented by higher order logics, but the representation becomes quite complex mathematically. Knowledge representation has to be flexible and allow for many exceptions. Knowledge has to continue to evolve as it is being accumulated, and the representation scheme has to allow for easy modification.

The ontology web language (OWL) (McGuinness and Van Harmelen, 2004), is another knowledge representation scheme. The basic structure underlying OWL is the subject–predicate–object model that makes assertions about a resource. It extends the resource description framework (RDF) to enable comprehensive ontology representation, such as the definition and properties of concepts, relationships among concepts, and constraints on the relationships between concepts and their properties. OWL, however, is best suited for declarative knowledge. It does not allow fuzzy knowledge and time dependencies in knowledge.

Networks are yet another knowledge representation alternative. All the variables together are represented as a directed acyclic graph whereby arcs are causal connections between two variables and the truth of the former directly affects the truth of the latter. For example, Bayesian networks (Pearl, 1985) (also known as belief networks) provide a means of expressing the joint probability distributions over many interrelated hypotheses. Another type of network representation is the semantic network (Sowa, 1987). Such networks represent knowledge in the form of interconnected nodes and arcs. Semantic networks can be definitional and assertional. Definitional networks focus on *IsATypeOf* relationships between a concept and a newly defined sub-type. Sub-types inherit properties of a super-type. The assertional network asserts propositions that are conditionally true. They are generally evaluated based on common sense.

The type of knowledge representation scheme adopted is a function of the type of knowledge to be represented. We believe an expanded form of semantic networks using more than definitional and assertional relationships offers the most flexible and effective scheme. We explain such a scheme in the next section.

To accelerate the discovery of knowledge, several generic knowledge databases have been created for commonly understood knowledge. Among the commonly used knowledge databases are (1) Cyc (Lenat and Guha, 1989), a logic-based repository of common-sense knowledge; (2) WordNet (Fellbaum, 1998), a widely used universal database of word senses; and (3) the Open Mind Common Sense project (Singh, 2002), a second-generation commonsense database. The Common Sense project crowd sources its knowledge through volunteers. Such knowledge stores have already been immensely helpful to a large number of semantic application development efforts.

3.7 A DEEP LEARNING FRAMEWORK FOR LEARNING AND INFERENCE

While NLP research has made great strides in producing artificially intelligent behaviors, none of the platforms (e.g., Google, IBM's Watson, and Apple's Siri) actually understand what they are doing – making them no different from a parrot that learns to repeat words without any clear understanding of what it is saying. Today, even the most popular NLP technologies view text analysis as a word- or pattern-matching task.

We believe that current deep learning approaches, some of which were reviewed in the previous section, are largely symptomatic and do not capture the real functioning of the human brain or logical inference. The two biggest gaps are the absence of contextual knowledge and a more realistic functional architecture.

The framework we propose has its roots in the extensive literature on cognitive architectures (see Langley et al., 2009, for an excellent review of this literature). Let's revisit the development process of the human brain. We acquire knowledge over time through our experiences, from the people around us, and from the structured education we may go through. We learn what is good and bad and all the shades of gray for all the contexts that we learn. And as the popular saying goes, "learning never stops."

It is well established in neuroscience research that the brain integrates multiple sources of information during learning through multiple modalities (Mousavi, Low, and Sweller, 1995). It appears that multisensory information processing is part and parcel of object perception and recognition in daily life, whereby the brain integrates the information from different modalities into a coherent percept (Ghazanfar and Schroeder, 2006). The human brain has evolved to develop, learn, and operate optimally in multisensory environments (Figure 3.19).

Figure 3.19 The human brain and multiple sources of input

As it relates to natural language processing, we need to recognize that language is a relatively recent cultural phenomenon and evolutionarily brain mechanisms for processing written script do not exist. The human brain has had to adapt to develop the ability to recognize the letter–speech association. In fact, Blomert and Froyen (2010) point out that in the last decade, neuroimaging studies have identified a brain region that shows specialization for fast visual word recognition, which is to say, the putative visual word form area (McCandliss et al., 2000) in the brain. Since fluency and automaticity are the most salient features of experienced reading (and from our perspective, interpretation), it is indeed plausible that a neural network involved in visual object recognition has specialized for recognizing visual letters and word forms in the human brain (McCandliss et al., 2003).

Years after children first learn to decode words into letters, a form of perceptual expertise emerges in which groups of letters are rapidly and effortlessly conjoined into integrated visual percepts, a process that is crucial to fluent reading ability. We need years of explicit instruction and practice before we start to exhibit any fluency in visual word recognition. This contrasts sharply with the way we learn to master spoken language. Infants and young children start to pick up and develop the many complexities of spoken language without explicit instructions at a time when literacy instruction is still far in the future. Recent electrophysiological evidence showed that it takes several years of reading instruction and practice before the first signs of automatic integration of letters and speech sounds appear in normally developing children. Letter–speech sound associations are cultural interventions and therefore biologically arbitrary in nature.

There are several key implications. First, through years of learning, we have learned numerous concepts, objects, and their attributes in different contexts. We have learned to recognize that the same attribute attached to the same object may evoke a different perception or recognition in us. Second, and a key implication, is that our knowledge is visible to us. While we don't know many things, what we do know is very deterministically known to us.

We also know from neuroscience that knowledge is organized in a hierarchical fashion. We may have an understanding of the more abstract but not the very granular knowledge in many situations. We may know a lot of concepts and topics but in a very shallow sense. We may know a few concepts and topics very deeply. In areas where we have acquired extensive knowledge, we will have a good understanding of all the granular objects, concepts, and attributes.

To model this cognitive behavior, we need to explicitly allow for extensive knowledge acquisition and representation in the architecture just like the brain does. The brain does face many fuzzy situations where it lacks precise knowledge to understand the information without ambiguity. In such situations, the brain resorts to the next higher level of structure in terms of the object and attempts to decide on that basis. When it has no information whatsoever, we end up with "I don't know."

Our stored knowledge is adjusted continually as we see the outcomes of our choices and actions. When I set an aggressive timeline for an action and it does not happen, and this happens a few times, I adjust my stored knowledge with that experience. At some point for similar projects, my stored knowledge begins to change my reaction in terms of setting aggressive timelines. I start setting conservative timelines from that point.

If the knowledge is clear and has no ambiguity, then our decision is very crisp. When it is not, then we take our best shot. However, we are always trying to acquire more specific and granular knowledge in areas that we need to, in order to make our knowledge more complete and decisions less ambiguous. If the decision involves the future, we try to ascertain our current views on the causal factors in our stored knowledge and forecast the future.

RAGE AI™ implements an enhanced semantic network based knowledge representation scheme arising from the rhetorical structure theory (Mann and Thomson, 1987).

RAGE AI™ aggregates, extracts from, classifies, and interprets text with respect to a given context in a supervised or unsupervised mode (Figure 3.20).

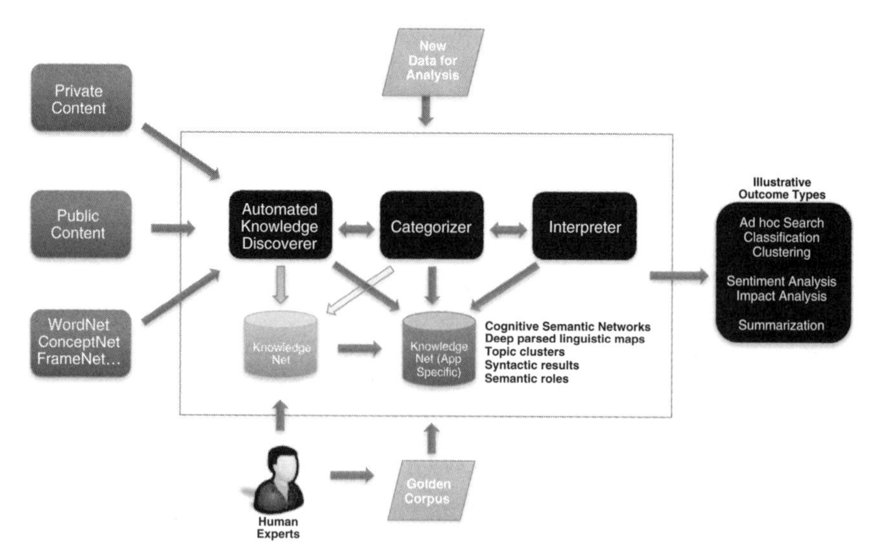

Figure 3.20 RAGE AI™: A deep learning framework for natural language understanding

It can be effectively deployed in any problem that requires search, classification or interpretation. Specifically, the high-level modules in RAGE AI™ include:

- *Knowledge discoverer* (KD), which automatically searches for knowledge related to the topic or entity of interest. Such knowledge will relate to all types of knowledge pertaining to the topic. KD can be applied to any type of public or private content.
- *KnowledgeNet* (KN), which stores domain-specific knowledge. Knowledge can be automatically discovered by KD and/or provided by experts. Experts can also edit the auto discovered knowledge if needed. KN stores knowledge using the extended semantic network referred to earlier.
- *Categorizer*, to classify the document into one or more topics. The categorizer naturally clusters the document into its constituent topics by examining the document and its discourse structure. It also overlays an external topic-specific semantic network for classification.
- *Impact analyzer*, which assesses the sentiment/potential impact of document on a topic of interest, including the analysis of sentiments of specific stakeholders that may be expressed in the document.

Interpretations of documents can be quite different depending on the context. Our goal is to automate the interpretation of the meaning of documents with reference to context, leveraging both assisted and unassisted knowledge discovery. Such an approach, which we term "impact analysis," allows us to create applications that can be used for decision making.

We provide a high level architectural overview of RAGE AI™ in Figure 3.21. As the figure shows, the framework consists of steps through which deep learning and understanding of natural language text are attained.

The content to be analyzed can be aggregated from a variety of sources including Internet URLs; micro-blogging sources like Facebook, Twitter, and Tumblr; internal documents; and content from premium sources. The framework provides the flexibility of adding new sources (URLs, Twitter users, internal documents, etc.) on the fly.

The semantic information extraction engine facilitates the extraction of specific types of content from the document, such as financial data from a financial statement, tables from a research report, and terms from a legal contract. RAGE AI™ incorporates a Domain Discourse Model to enable the creation and storage of domain-specific structural properties. For example, a news story generally has a title and a body; a research report might have a

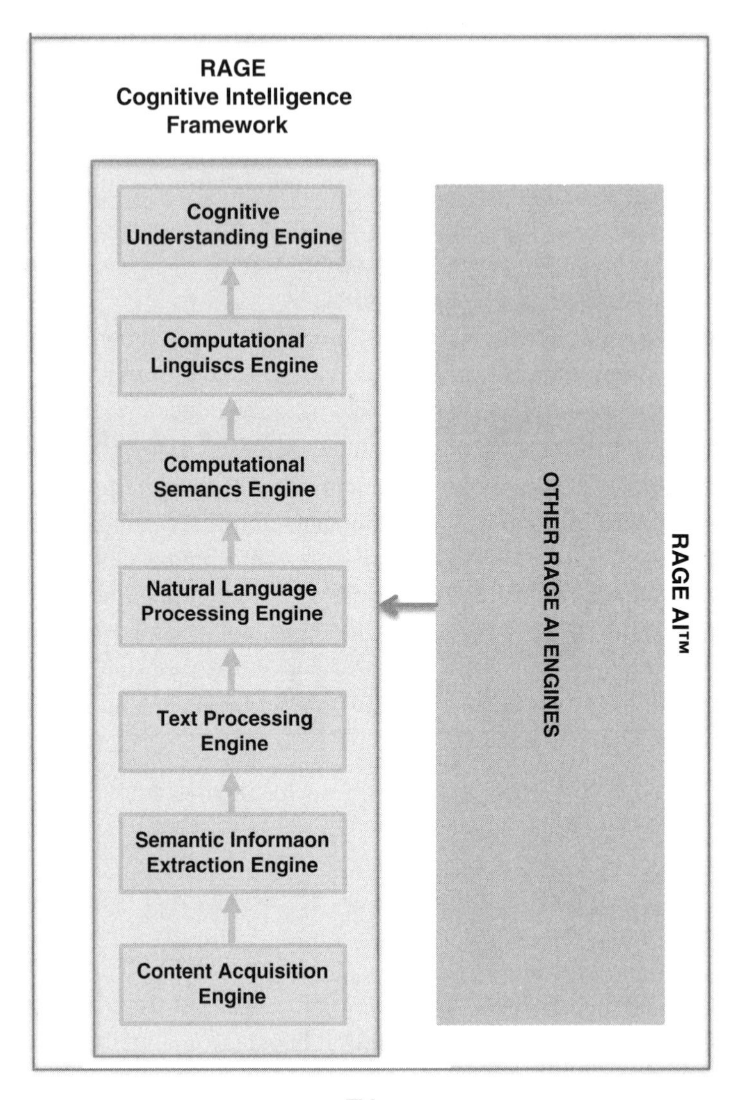

Figure 3.21 RAGE AI™ deep learning architecture

title, divided into sections, contain references and appendixes; a legal contract might have hierarchical structuring of sections and appendixes and exhibits; and an academic paper typically has an abstract, body, and references. Additional structural properties might include any other structural property that can aid the machine to improve its processing of the text; for example, news stories typically contain the main idea of the article toward the top of the news story and the rest is largely elaboration.

The text processing and natural language processing engines do syntactic and shallow parsing of the text. These steps clean up the text, tokenize, identify the part-of-speech roles that different words play, and identify named entities. The computational semantics engine begins the process of deep learning. It attempts to normalize concepts, disambiguate word senses, recalibrate parts-of-speech roles, disambiguate concepts based on the KnowledgeNet (KN), and generate dependency trees. Disambiguation of word senses and concepts is a function of the stored knowledge, and while this engine may not be 100% accurate in all the instances, it will approach 100% accuracy over time.

The computational linguistics engine (CLE) detects clauses and clause boundaries, semantic roles of the various words/chunks of text, and co-references across the text; it then clusters the document, and creates a document map (or graph). The output at this stage has a deep linguistic representation of the text document.

Finally, the cognitive understanding engine (CUE) attempts to mimic the human understanding of the document based on the output from the previous steps, and most important the output of CLE. CUE can classify the document if that is the desired outcome and/or interpret the document with respect to a defined topic, integrating the KnowledgeNet, if relevant. Expert knowledge, if any, may also be stored in the KnowledgeNet. Interpretation can be purely local, or global with the integration of KnowledgeNet and potentially expert assisted if the expert has provided their knowledge to the KnowledgeNet.

3.7.1 Conceptual Semantic Network

A key central aspect of RAGE AI™ is the conceptual semantic network (CSE). CSE builds on all the other deep parsing steps outlined in Figure 3.19 and is one of the outputs of the cognitive understanding engine. The CSE is an exhaustive enumeration of relationships between any two topics. It enumerates all possible relationships between two concepts in addition to the typical "definitional" relationships (synonym, hypernym, meronym) characteristic of a typical ontology. It has its roots in rhetorical structure theory, which is a widely accepted framework for discourse analysis (Mann and Thomson 1987, 1988; Soricut and Marcu, 2003; Reitter, 2003; LeThanh et al., 2004; Baldridge and Lascarides, 2005; Subba and Di Eugenio, 2009; Sagae, 2009; Hernault et al., 2010).

Prior work on automated relationship extraction was limited to local relationships like hypernymy and meronymy (Olga Acosta et al., 2012; Inga

Gheorgita et al., 2012; Anthony Fader, 2012). We extend such work significantly by extracting a global set of relationships and combining such relationship extraction with a Bayesian algorithm in order to compute and rank the relative strengths of such relationships.

To discover relationships and related concepts, we expand the idea that coherent text in discourse implies a set of structural relationships between related concepts and grounded in language. Relationships are identified between two *concepts* or two *events* or a *concept* and an *event,* and vice-versa. RAGE AI™ also implements a *compound concept identifier* and *concept normalizer*, that intelligently reduces complex noun-phrases into specific *normalized concepts*, so that different relations about the same event or concept can be identified or inferred. An *inference engine* within the COD is then able to perform several contextual inferences to create a multilevel, hierarchical, conceptual semantic network.

Some of the relations that the CSE tries to find are the following:

1. *Attribution* Named entity A is expressing something about concept B (e.g., *France* said that it will back Palestine on its *non-member observer entity* status).
2. *Causal* Event_A causes event_B (e.g., the stagnant housing industry got a rare boost last month, as more people bought new homes after the worst winter for sales in almost 50 years).
3. *Comparison* Event_A is compared to event_B (e.g., the housing sector continues to lag, whereas other sectors have begun a rebound in earnest).
4. *Conclusion* Event_A is a conclusion of event_B (e.g., the inflation_rate over the longer run has primarily determined monetary_policy and hence the committee has the ability to specify a longer run goal for inflation).
5. *Conditional* If Event_A, then event_B (e.g., if home prices dip again, then consumers may curb their spending).
6. *Contrast* Event_A and event_B have contrasting behaviors.
7. *Contra-expectation* Event_A occurs even when event_B occurred, which was opposite to the expectations (e.g., the housing_market continues to remain low, though it did get a significant boost in March).
8. *Elaboration* Event_A is a elaboration of event_B (e.g., economists forecast that incomes may also rise).
9. *Hypernym* Event_A is a hypernym of event B (e.g., retailers such as Home Depot Inc.).

10. *Justification* Concept_B is used to justify the event on concept A.
11. *Reason* Event_A is why there occurred event_B (e.g., pending_home_sales are considered a leading_indicator because they track contract signings).
12. *Result* Event_A is a result of event_B (e.g., rising incomes in the respective foreign countries support increased sales).
13. *Temporal_Simultaneous* Event_A occurred simultaneously with event_B (e.g., in Bristol, sales dropped 43.8% in April compared_with the same month last year, while the median sales price fell 3% to $225,000).
14. *Temporal_Succession* Event_A is succeeded by event_B (e.g., many markets began a decline, *once* those tax credits expired in April).

Examples

"Consumer Confidence in the United States fell last week to the lowest level since August as rising prices squeeze household budgets."

Clause 1: *Consumer Confidence in the United States fell last week to the lowest level since August*
Clause 2: *as rising prices squeeze household budgets*
Relation 1: [*rising prices*] **CAUSE** [*household budgets*]
Relation 2: [*rising prices*] **CAUSE an effect on** [*household budgets*]
Relation 3: (Derived) [household budgets] **CAUSES an effect on** [consumer confidence]

Optionally, we derive a confidence score with each relationship. The confidence score is directly proportional to the evidence of a specific relationship in our corpus. With every new relationship that is identified from the text, an adaptive score is given to the new relationship and all the evidence of this relationship in our database. Our approach to finding the existence of a relationship benefits from our ability to create normalized concepts which makes it independent of part of speech.

3.7.2 Knowledge Discoverer

As stated before, we believe "domain knowledge" and knowledge acquisition is a key missing link in the majority of current machine learning methods and applications. At the same time, knowledge acquisition can be a

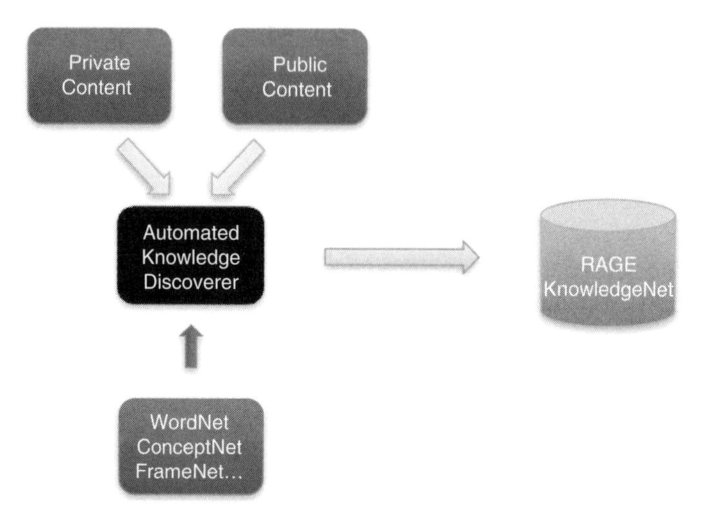

Figure 3.22 Automated knowledge discoverer architecture

time-consuming task and thus has deterred many from attempting it. In the context of natural language, RAGE AI™ dramatically speeds up knowledge acquisition through its knowledge discoverer (KD). Given a topic of interest, KD automatically discovers related concepts and their relationships to the topic of interest in a target corpus. The target corpus can be public and/or private content. Experts can also provide knowledge or edit the acquired knowledge if desired. This interaction can be iterative. Thus KD can function in a completely unsupervised, semi-supervised, or fully supervised mode.

KD works in conjunction with KnowledgeNet. KD first looks in KnowledgeNet to retrieve any stored knowledge on the concept of interest. The results are then optionally aggregated with a fresh interrogation of public and/or private content to create the most updated knowledge base on the concept. Such updated knowledge is then available to the rest of the RAGE AI™ framework like the classifier and impact analyzer (Figure 3.22).

3.7.3 Computational Linguistics Engine

The computational linguistics engine (CLE) draws on the output of all the components that precede it, as shown in Figure 3.21. CLE parses the output of the previous steps and the text using linguistic concepts and rules to produce a deeply parsed interpretable version of the text. This includes detecting clauses and clause boundaries, semantic roles, co-reference resolution, and the clustering of the text either according to its local organization (unsupervised) or according to stored knowledge or query (semi-supervised).

CLE implements both anaphoric and cataphoric referential relationships. While most co-reference resolution models use a single function over multiple mentions in a local context, it is well understood that such local resolution will yield poor performance (Bengston and Roth, 2008; Finkel and Manning, 2008; Stoyanov, 2010). Raghunathan et al. (2010) propose a multi-pass sieve that applies tiers of deterministic co-reference models one at a time from highest to lowest precision.

CLE's co-reference resolution method is a flexible, multi-pass algorithm. In addition to deterministic co-reference resolution rules, our approach allows for the implementation of multiple contextual constraints and declarative pragmatic relationships relevant to the context.

Our approach to identification of "topics" associated with a document involves the finding of natural conceptual clusters in the document. Such clusters are found by identifying the semantic similarities between all sentences and paragraphs in the document. Such semantic similarity includes co-referential relationships, conceptual relationships, and relevant relationships in KnowledgeNet.

Figure 3.23 shows a document decomposed into cluster 1 and cluster 2 based on the linguistic relationships between sentences in the document. Cluster 1 was found as a cluster of sentences S_1, S_2, S_3, and S_{10}, with the thickness of the relationship arrows depicting the strengths of their linguistics relationships (or semantic similarities).

The strength of semantic similarity between two sentences, S_a (with M noun-phrases), S_b (with N noun-phrases) in a document is represented by $S(a, b)$:

$$S(a,b) = \sum_{i=1}^{M} \sum_{j=0}^{N} \sum_{k=0}^{K} w_k f_k(x_i, x_j)$$

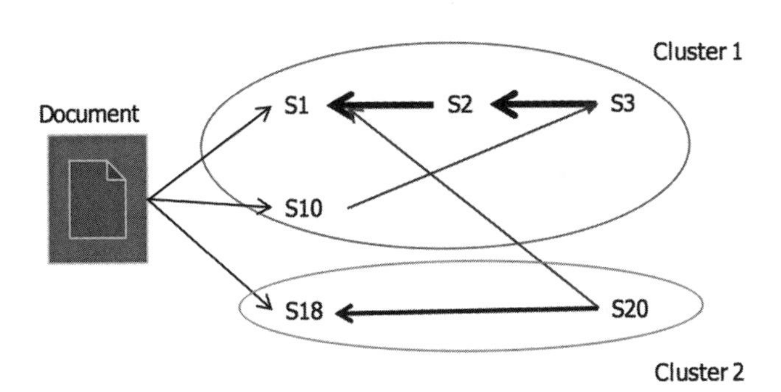

Figure 3.23 Breaking a document into sub-graphs

A threshold (T) denoting the strength of semantic similarity is used to instruct the computer to disregard weak semantic matches. T can be defined using a supervised algorithm or simply using an empirically derived value. Use of a threshold for $S(a, b)$ obtains

$$S(a,b) = S(a,b),\ \mathit{if}\ S(a,b) > T$$
$$S(a,b) = 0,\ \mathit{if}\ S(a,b) < T$$

Each cluster, C, is associated with a primary topic, T, and secondary topics ($ST_1 \ldots ST_n$). The subject of the primary sentence of the cluster is chosen as the primary topic of the cluster, C.

The CLE generates a document map as an acyclic (if not considering cataphora) directed graph, with the following characteristics:

1. Each node is a sentence in the document.
2. Each edge represents the semantic similarity strength between two sentences. Zero strength edges are not considered as edges.
3. The graph may consist of several cliques, resulting in several subgraphs.

Given a root node in the document map, if we follow any one edge of the root node, we would be following one of the major themes of the document. Each clique in the document map is termed as a cluster of semantically relevant sentences. A cluster is considered as a set of sentences falling in the same clique. Strength of a cluster C with sentences $s_0, s_1, \ldots, s_L$ is defined by

$$C(s_{0,} s_{1,} \cdots S_L) = \sum_{j=0}^{L} \sum_{i=0}^{L} S(s_i, s_j)$$

For unsupervised categorization, we can easily identify the central topic of the article by identifying the cluster that has the maximum strength.

Depending on the type of document, there may be a single or many clusters (or discourse units). Beyond intrinsic characteristics of the document based on its objective, many clusters in a document may indicate the use of pragmatic concepts, such as “xerox” for “copy.” CLE is able to minimize such unknown pragmatic relationships in two ways. First, there is likelihood that the pragmatic relation is already discovered and stored in KnowledgeNet. CLE will look in KnowledgeNet to see if there are pragmatic relationships. Second, KnowledgeNet provides a facility for storing known domain-specific pragmatic relationships.

In a semi-supervised learning mode, expert curated knowledge networks can be overlaid on the document clusters. The sentence in which a topic from an expert-provided knowledge network is found becomes the main sentence and the primary topic for the cluster. CLE allows for the specification of domain-specific discourse properties in determining the strength of relevance of a document to a topic. For example, users can prescribe different weights for different parts of the document based on location, and/or based on the topic's semantic role in the sentence (subject, object, etc.).

3.7.4 Impact Analysis

To analyze the impact of a document on a topic/entity of interest, the CUE has an elaborate, flexible, and powerful framework that extracts *impact phrases* from documents and interprets them in a normalized fashion with respect to the topic/entity of interest. Impact phrases are much more elaborate in a "meaning" sense than sentiment lexicons used for sentiment analysis. They require a complete linguistic analysis of the discourse and identification of the language structure. For example, the RAGE platform identifies such things as the subject of the sentence, whether it is in active or passive voice, whether the verb denotes impact or the verb is an impact-neutral transitive verb, and so on.

Impact phrases are any set of words that communicate impact; they can be verbs, nouns, adjectives, and adverbs. Impact phrases are an aggregation of primary impact words and words that accentuate the impact; adjectives and adverbs are generally used to accentuate impact; we refer to them as "impact intensity."

Besides primary impact phrases, the system extracts descriptive attributes like time, quantity, and location. For example, in "Housing starts rose significantly by 3% in January," the verb "rose" is the primary impact word, the adverb "significantly" describes intensity, the "3%" is also reflective of intensity, and "January" is an indicator of time. Such phrases can contain "negation," for example, "Housing starts did not rise in January." Directional reversal can sometimes occur in a more subtle manner; it can also manifest by the use of "anti" words like "anti-immigration" or "unemployment."

To achieve consistency and comparison across topics, we normalize the impact assessment to a discrete set of outcomes and associated scores; currently, we use five discrete categories: significantly positive, positive, neutral, negative, and significantly negative. The same impact phrase can have completely opposite interpretation for two topics. For example, "increase in oil prices" will have a +ve impact on oil companies and −ve impact on transportation companies. We enable the configuration of impact direction at a topic/normalized impact level.

3.7.5 Formulation of the Impact Analysis Problem

Consider a document D_j tagged by topics $T_1, T_2, \ldots, T_N$. The document is further divided into several sub-documents (clauses and sentences; decided dynamically), $D_{j1}, D_{j2}, \ldots, D_{jM}$. Each sub-document has been analyzed and assigned impacts $I_j = I_{j1}, I_{j2}, \ldots, I_{jM}$. Then the document-level score of the document with respect to each topic T_j is written as

$$Score\,(T_i, D_j) = (1/Z) \sum_{k=0}^{K} w_{jk} * I_{jk}$$

where w_{jk} is the weight of the impact, as a function of the sub-document D_{jk}, and Z is the normalization factor.

The I_{jk} are found by the impact analyzer linguistically, using the impact phrases and amplifying them according to the corresponding impact and intensity phrases (including locations, time periods, and quantities). The weight of the impact is the function of the sub-document's cluster and positional information. In the simplest implementation, the w_{jk} is 1 when the sub-document D_{jk} belongs to the strongest cluster for that topic and 0 otherwise. Even within the cluster, the w_{jk} can vary based on the position of the sub-document within the strongest cluster. We have several scoring schemes that match various theories of discourse analysis.

This kind of a scoring scheme assigns a N-dimensional score (for N topics in the system/ontology) to a document. This, when viewed as a time series, provides an assessment of the topic/event or entity as a function of time.

3.8 SUMMARY

There is a lot of interest in timely analysis and interpretation of the massive amounts of unstructured data and structured data that are being created in the new digital age. In this chapter, we concentrated on insight and intelligence. We had discussed efficiency in Chapter 2. We began by reviewing the relationship between information and decision making. We discussed at length the motivation behind information asymmetry as a source of value. We discussed taxonomic organization and AI and machine learning. There is a lot of mystique surrounding machine learning. In particular, different people have very different understandings of what machine learning is capable of and what it is not capable of. We hope now the reader has a better sense of how the methods and concepts of machine learning fit in the overall scheme. We also describe briefly the most popular machine learning methods.

We conclude the chapter with a description of RAGE AI™, which is a traceable deep learning cognitive intelligence framework. We argue that whether we are dealing with structured or unstructured data or a mix of both, the rush to create machine learning solutions has so far centered on computational statistics. Machine learning based on computational statistics can be effective with homogeneous data sets, but will still require a significantly large corpus for training. However, machine learning solutions that are being developed using traceable approaches like RAGE AI™ are likely to be more trusted and will require smaller corpus for learning.

RAGE AI™ has been validated in numerous enterprises across a diverse set of business domains:

- *Business viability risk* The platform is used for continuous assessment of business risk, where an intelligent machine continuously assesses whether developments around the world can potentially impact the viability of a customer or supplier or partner.
- *Finding alpha in financial markets* An RAGE AI™ powered application continuously assesses the impact of news, blog opinions, research, and social media on a company's intrinsic value. These signals enable asset managers to make investment decisions and exploit market inefficiencies. (This application is described in detail in Chapter 6.)
- *Customer satisfaction* To assess the true nature of customer complaints and for an intelligent, timely, and consistent response to the customer, RAGE AI™ analyzes millions of customer complaints along with the associated interactions between client service representatives and customers to develop continuous insight and intelligence into the root cause of issues and the quality and accuracy of the responses. The overall goal is to focus on the real issues and keep customer satisfaction at the highest levels.
- *Cognitive help desk* Similar to the customer complaint application, an intelligent application to understand help desk requests and respond intelligently and in a timely manner. The machine is trained on all the available material on a company's systems, policies and practices.
- *Customer, market, and competitive intelligence* A RAGE AI™ application surfaces insight and emerging trends across all business sectors

and industries on a continuous basis to provide consultants and marketing professionals insight into emerging trends, topics of interest, and competitive behavior. Users can then design their appropriate responses.

- *Sales lead generation* A RAGE AI™ application interprets a vast amount of structured and unstructured content to identify potential sales leads. Company experts can provide their expertise to RAGE AI™ applications to tailor them to their own needs. Firms typically manually analyze news and other public information in order to identify potentially qualified leads. We found in one company that the application generated 200% more relevant leads than the human team, and missing only 5–10%.
- *Wealth management advisory* The wealth management advisory market is in the throes of a major transformation. Historically, this market is very hands on, led by financial advisors who use their experience and knowledge to advice their clients on wealth management. A RAGE AI™ application is used by leading wealth management firms to intelligently automate the end to end process across the wealth spectrum – from retail to high net worth and institutional clients. (This case study is discussed in more detail in Chapter 5.)
- *Efficacy of prescription drugs* There is a lot of useful information that is posted regarding the effects of drugs, and there is a steady stream of high-quality scientific research that is being published continuously. The FDA also publishes quarterly statistics on adverse effects. A RAGE AI™ application synthesizes all this information and attempts to identify trends and insight that should be of interest to consumers, pharma companies, insurers, and regulators. We found that in many cases the RAGE AI™ application could have easily predicted FDA action, such as black box warnings, well before the action finally happened.
- *Reconciliation* RAGE AI™ can reconcile transactions across multiple documents sometimes from multiple systems. For example, a large food manufacturer and distributor needs to reconcile its planned transportation routes and costs with the transportation costs based on actual routes and to triangulate costs in contracts with the transportation vendors. In another example, a global financial services firm needed to reconcile invoices to its customers with the underlying contracts. RAGE AI™ applications provide real time reconciliation in these cases.

These use cases go across many industries – financial services, logistics, food and beverage, life sciences, and consulting services. The machine intelligence in the above-mentioned cases was acquired largely in an automated fashion and also benefited from expert knowledge provided by human experts.

In all these cases, the readers should take away the fact that these are real, functioning applications that are delivering insight and intelligence to enterprises today. This is not hypothetical. The level of insight will continue to advance in each of these areas as the machines get more experiential data and the experts provide more knowledge. Another very important take away for the reader is that machine learning does not have to be a black box. Deep learning approaches can be completely contextual and traceable.

REFERENCES

Acosta, O., Aguilar, C., and Sierra, G. (2010). A method for extracting hyponymy–hypernymy relations from specialized corpora using Genus terms. In *Proceedings of the Workshop in Natural Language Processing and Web-based Technologies*. Córdoba, Argentina: Universidad Nacional de Córdoba, 1–10.

Alpaydim, E. (2010). *An Introduction to Machine Learning*, 2nd ed. Cambridge: MIT Press.

Baker, L. D., and McCallum, A. K. (1998). Distributional clustering of words for text classification. In *Proceedings of SIGIR'98, 21st ACM International Conference on Research and Development in Information Retrieval*. New York: ACM Press, 96–103.

Baldridge, J., and Lascarides, A. (2005). Probabilistic head-driven parsing for discourse structure. In *Proceedings of the Ninth Conference on Computational Natural Language Learning*. Stroudsburg, PA: Association for Computational Linguistics, 96–103.

Barwise, J. (1977). An introduction to first-order logic. *Handbook of Mathematical Logic*. (Studies in Logic and the Foundations of Mathematics). Amsterdam: North-Holland, 5–46.

Basili, R., Moschitti, A., and Pazienza, M. T. (2000). Language-sensitive text classification. In *Proceedings of RIAO'00, 6th International Conference "Recherche d'Information Assistee par Ordinateur*." Paris, France, 331–343.

Bell J. F. (1999). Tree-based methods. In *Machine Learning Methods for Ecological Applications*, edited by A. H. Fielding. Boston: Kluwer Academic, 89–105.

Bengio, Y. (2009). Learning deep architectures for AI. *Foundations and Trends in Machine Learning* 2(1): 1–127.

Bengston, E., and Roth, D. (2008). Understanding the value of features for coreference resolution. In *Proceedings of the 2008 Conference on Empirical Methods in*

Natural Language Processing, Honolulu, October. Stroudsburg, PA: Association for Computational Linguistics, 294–303.

Bishop C. M. (1995). *Neural Networks for Pattern Recognition*. Oxford, UK: Clarendon Press.

Blair-Goldensohn, S., Hannan, K., McDonad, R., Neylon, T., Reis, G. A., and Reynar, J. (2008). *Building a Sentiment Summarizer for Local Service Reviews*. WWW 2008 Workshop on NLP Challenges in the Information Explosion Era (NLPIX 2008) (2008), http://www.cl.cs.titech.ac.jp/~fujii/NLPIX2008/paper3.pdf.

Blomert, L., and Froyen, D. (2010). Multi-sensory learning and learning to read. *International Journal of Psychophysiology* 77(3): 195–204.

Breiman L., Friedman J., Olshen R. A., Stone, C. J. (1984). *Classification and Regression Trees*. Belmont, CA: Wadsworth International Group.

Breiman, L. (1996). Bagging predictors. *Machine Learning* 24: 123–140.

Breiman, L. (2001a). Random forests. *Machine Learning* 45: 5–32.

Breiman, L. (2001b). Statistical modeling: The two cultures. *Statistical Science* 16(3): 199–231.

Breiman, L. (2004). Consistency for a simple model of random forests. Technical Report 670. UC Berkeley.

Cade, M. (2006). Google AI takes historic match against Go Champ with third straight win. *Wired*, March 2.

Carenini, G., Ng, R., and Pauls, A. (2006). Multi-document summarization of evaluative text. In *Proceedings of the Conference of the European Chapter of the Association for Computational Linguistics (EACL)*. Stroudsburg, PA: Association for Computational Linguistics, 305–312.

Caropreso, M. F., Matwin, S., and Sebastiani, F (2001). A learner-independent evaluation of the usefulness of statistical phrases for automated text categorization. In Amita G. Chin, ed., *Text Databases and Document Management: Theory and Practice*. Hershey, PA: Idea Group Publishing, 78–102.

Cho, E., and Chon, T.-S. (2006). Application of wavelet analysis to ecological data. *Ecological Informatics* 1(3): 229 –233.

Cohen, L., Dehaene, S., Naccache, L., Lehéricy, S., Dehaene-Lambertz, G., Hénaff, M. A., and Michel, F. (2000). The visual word form area: Spatial and temporal characterization of an initial stage of reading in normal subjects and posterior split-brain patients. *Brain*, 123 (Pt 2): 291–307.

Cole, S., Royal, M., Valtorta, M., Huhns, M., and Bowles, J. (2006). A lightweight tool for automatically extracting causal relationships from text. In *Proceedings of the IEEE SoutheastCon,* IEEE, Tennessee. Washington, DC: IEEE, 125–129.

Coleman, A. (2006). *Dictionary of Psychology*, 2nd ed. New York: Oxford University Press, p. 688.

Crammer, K., and Singer, Y. (2001). On the algorithmic implementation of multiclass kernel-based vector machines. *Journal of Machine Learning Research* 2: 265–292.

Dave, K., Lawrence, S., and Pennock, D. M. (2003). Mining the peanut gallery: Opinion extraction and semantic classification of product reviews. In Proceedings of the 12th international conference on World Wide Web. New York: ACM, 519–528.

De'ath, G., and Fabricius, K. E. (2000). Classification and regression trees: A powerful yet simple technique for ecological data analysis. *Ecology* 81(11): 3178–3192.

di Stefano, T. F. (2006). Information asymmetry: Shattered by technology. *E-Commerce Times*, November 24.

Drake J. M., Randin C., and Guisan A. (2006). Modelling ecological niches with support vector machines. *Journal of Applied Ecology* 43(3): 424–432.

Duan, K. B., and Keerthi, S. S. (2005). Which is the best multiclass SVM method? An empirical study. *Multiple Classifier Systems. LNCS* 3541: 278–285.

Elith, J., Graham, C. H., Anderson, R. P., Dud'ık, M., Ferrier, S., Guisan, A., and Hijmans, R. J., et al. (2006). Novel methods improve prediction of species' distributions from occurrence data. *Ecography* 29(2): 129–151.

Eppler, M. J., and Mengis, J. (2004). The concept of information overload: A review of literature from organization science, accounting, marketing, MIS, and related disciplines. *Information Society* 20: 325–344.

Fader, A., Soderland, S., and Etzioni, O. (2011). Identifying relations for open information extraction. In *Proceedings of the 2011 Conference on Empirical Methods in Natural Language Processing*, Edinburgh, Scotland. Stroudsburg, PA: Association for Computational Linguistics, 1535–1545.

Fellbaum, C. (1998). *WordNet: An Electronic Lexical Database: Language, Speech, and Communication*. Cambridge: MIT Press.

Finkel, J., and Manning, C. (2008). Enforcing transitivity in coreference resolution. In *Proceedings of ACL-08: HLT, Short Papers (Companion Volume)*, Columbus, OH. Stroudsburg, PA: Association for Computational Linguistics, 45–48.

Forsbom, E. (2005). Rhetorical structure theory in natural language generation. Uppsala University and GSLT.

Fortune (2012). What data says about us. September 24, p. 163.

Friedmann, M. (1977). Consumer use of informational aids in supermarkets. *Journal of Consumer Affairs* 11(1): 78–155.

Frydman, H., Altman, E. I., and Kao, D. (1985). Introducing recursive partitioning for financial classification: The case of financial distress. *Journal of Finance* 40(1): 269–291.

Gamon, M., Aue, A., Corston-Oliver, S., and Ringger, E. (2005). Pulse: Mining customer opinions from free text. In *Proceedings of the 6th International Symposium on Intelligent Data Analysis (IDA)*, Madrid, Spain. Lecture Notes in Computer Science. Berlin: Springer, 121–132.

Gantz, J., and Reinsel, D. (2012). *The Digital Universe in 2020: Big Data, Bigger Digital Shadows, and Biggest Growth in the Far East*. Framingham, MA: IDC.

Geman, S., Bienenstock, E., and Doursat, R. (1992). Neural networks and the bias/variance dilemma. *Neural Computation* 4(1): 1–58.

Ghazanfar, A. A., and Schroeder, C. E. (2006). Is the neocortex essentially multisensory? *Trends in Cognitive Sciences* 10: 278–285.

Gheorghita, I., and Jean-Marie, P. (2012). Towards a methodology for automatic identification of hypernyms in the definitions of large-scale dictionary. In *Proceedings of the Eight International Conference on Language Resources and Evaluation (LREC'12)*. Istanbul: European Language Resources Distribution Agency, 2614–2618.

Girju, R., and Moldovan, D. (2002). Text mining for causal relations. In *Proceedings of FLAIRS-02*. Palo Alto, CA: American Association for Artificial Intelligence, 360–364.

Godbole, N., Srinivasaiah, M., and Skiena, S. (2007). Large-scale sentiment analysis for news and blogs. In *Proceedings of the International Conference on Weblogs and Social Media (ICWSM)*, Boulder, CO.

Grise, M., and Gallupe, R. B. (1999/2000). Information overload: Addressing the productivity paradox in face-to-face electronic meetings. *Journal of Management Information Systems* 16(3): 157–185.

Guisan, A., and Zimmermann, N. E. (2000). Predictive habitat distribution models in ecology. *Ecological Modelling* 135(2–3): 147–186.

Hansen, M. T., and Haas, M. R. (2001). Competing for attention in knowledge markets: Electronic document dissemination in a management consulting company. *Administrative Science Quarterly* 46: 1–28.

Hastie, T., Tibshirani, R., and Friedman, J. H. (2001). *The Elements of Statistical Learning: Data Mining, Inference, and Prediction*. New York: Springer.

Hatzivassiloglou, V., and McKeown, K. R. (1997). Predicting the semantic orientation of adjectives. In *Proceedings of the 8th Conference of European Chapter of the Association of Computational Linguistics*. Stroudsburg, PA: Association for Computational Linguistics, 174–181.

Hernault, H., Prendinger, H., duVerle, D. A., and Ishizuka, M. (2010). HILDA: A discourse parser using support vector machine classification. *Dialogue and Discourse* 1(3): 1–33.

Hogeweg, P. (1988). Cellular automata as a paradigm for ecological modeling. *Applied Mathematics and Computation* 27(1): 81–100.

Hopfield, J. J. (1982). Neural networks and physical systems with emergent collective computational abilities. In *Proceedings of the National Academy of Sci- ences, USA*, 79(8): 2554–2558.

Hsu, C, and Lin, C. (2002). A comparison of methods for multiclass support vector machines. *IEEE Transactions on Neural Networks* 13(2): 415–425.

Hu, M., and Liu, B. (2004). Mining and summarizing customer reviews. In *Proceedings of the Tenth ACM SIGKDD International Conference on Knowledge Discovery and Data Mining*. New York: ACM, 168–177.

Huong, L., Abeysinghe, G., and Huyck, C. (2004). Generating discourse structures for written texts. In *Proceedings of the 20th International Conference on Computational Linguistics*. Stroudsburg, PA: Association for Computational Linguistics, 329–335.

Hunt, R. E., and Newman, R. G. (1997). Medical knowledge overload: A disturbing trend for physicians. *Health Care Management Review* 22: 70–75.

Hutter, M. (2005). *Universal Artificial Intelligence*. Berlin: Springer.

Jacobs, P. J. (1992). Joining statistics with NLP for text categorization. In *Proceedings of the Third Conference on Applied Natural Language Processing*, Trento, Italy. Stroudsburg, PA: Association for Computational Linguistics, 178–185.

Jacoby, J., Speller, D. E., and Berning, C. K. (1974). Brand choice behavior as a function of information load: Replication and extension. *Journal of Consumer Research* 1: 33–43.

Sparck Jones, K. (1972). A statistical interpretation of term specificity and its application in retrieval. *Journal of Documentation* 28: 11–21.

Kim, S. M., and Hovy, E. (2004). Determining the sentiment of opinions. In *Proceedings of the COLING/ACL Main Conference Poster Sessions*. Stroudsburg, PA: Association for Computational Linguistics, 611–618.

Kim, S. M., and Hovy, E. (2006). Automatic identification of pro and con reasons in online reviews. In *Proceedings of the COLING/ACL Main Conference Poster Sessions*, 483–490.

Kohonen, T. (2001). *Self-Organizing Maps*. Berlin: Springer.

Kudo, T., and Matsumoto, Y. (2004). A boosting algorithm for classification of semi-structured text. In *Proceedings of the Conference on Empirical Methods in Natural Language Processing (EMNLP)*. Barcelona, Spain. Stroudsburg, PA: Association for Computational Linguistics, 301–308.

Langley, P., Laird, John, E., and Rogers, S. (2009). Cognitive architectures: Research issues and challenges. *Cognitive Systems Research* 10(2): 141–160.

Lecun, Y., Bottou, L., Bengio, Y., and Haffner, P. (1998). Gradient-based learning applied to document recognition. In *Proceedings of the IEEE*, 86(11): 2278–2324.

Lek, S., Delacoste, M., Baran, P., Dimopoulos, I., Lauga, J., and Aulagnier, S. (1996). Application of neural networks to modelling nonlinear relationships in ecology. *Ecological Modelling* 90(1): 39–52.

Lenat, D., and Guha, R. (1989) *Building Large Knowledge-Based Systems: Representation and Inference in the Cyc Project*. North Reading, MA: Addison-Wesley.

Lloyd, L., Kechagias, D., Skiena, S. (2005). Lydia: A System for large-scale news analysis. In *String Processing and Information Retrieval (SPIRE 2005)*. Lecture Notes in Computer Science. Springer Berlin, 3772: 161–166.

Libowski, Z. (1975). Sensory and information inputs overload: Behavioral effects. *Comprehensive Psychiatry* 16: 199–221.

Lloyd, L., Kechagias, D., and Skiena, S. (2006). Identifying co-referential names across large corpora. In *Combinatorial Pattern Matching (CPM 2006)*. Lecture Notes in Computer Science. Berlin: Springer, 4009: 12–23.

Lodhi, H., Saunders, C., Shawe-Taylor, J., Cristianini, N., and Watkins, C. (2002). Text classification using string kernels. *Journal of Machine Learning Research* 2: 419–444.

Luger, G., and Stubblefield, W. (2004). *Artificial Intelligence: Structures and Strategies for Complex Problem Solving*, 5th ed. San Francisco: Benjamin Cummings.

Luhn, H. P. (1957). A statistical approach to mechanized encoding and searching of literary information. *IBM Journal of Research and Development* 1 (4): 315.

Mann, W. C., and Thompson, S. A. (1987). Rhetorical structure theory: A framework for the analysis of texts. *IRPA Papers in Pragmatics.* Antwerp, Belgium: International Pragmatics Association, 1(1): 79–105.

Marcus, G. (2012). Is "deep learning" a revolution in artificial intelligence? *The New Yorker*, November, 25.

Marsden, P (2013). Fast facts: Information overload 2013. *Digital Intelligence*, November 14.

McCandliss, B. D., Posner, M. I., and Givon, T. (1997). Brain plasticity in learning visual words. *Cognitive Psychology* 33: 88–110.

McCandliss, B. D., Cohen, L., and Dehaene, S. (2003). The visual word form area: Expertise for reading in the fusiform gyrus. *Trends in Cognitive Sciences* 7(7): 293–299.

McCarthy, J. (2007). What is artificial intelligence? Computer Science Department, Stanford University. http://www-formal.stanford.edu/jmc/whatisai/.

McDonald, R., Hannan, K., Neylon, T., Wells, M., and Reynar, J. (2007). Structured models for fine-to-coarse sentiment analysis. In *Proceedings of the Annual Conference of the Association for Computational Linguistics (ACL)*. Stroudsburg, PA: Association for Computational Linguistics, 432–439.

McGuinness, D., and Van Harmelen, F. (2004). OWL web ontology language overview. *W3C recommendation.*

Meier, R. L. (1963). Communications overload: Proposals from the study of a university library. *Administrative Science Quarterly* 7: 521– 544.

Miller, G.A. (1995). Wordnet: A lexical database. *Communications of the ACM* 38(11): 39041.

Mitchell, T. (2006). The discipline of machine learning. School of Computer Science. Carnegie-Mellon University. CMU-ML-06-108, July.

Mousavi, S. Y., Low, R., and Sweller, J. (1995). Reducing cognitive load by mixing auditory and visual presentation modes. *Journal of Educational Psychology* 87(2): 317–334.

Mullen, T., and Collier N. (2004). Sentiment analysis using support vector machines with diverse information sources. In *Proceedings of the Conference on Empirical Methods in Natural Language Processing (EMNLP)*, July. Stroudsburg, PA: Association for Computational Linguistics, 412–418.

Nasri, G. (2013). Solving information asymmetry: How today's companies are empowering consumers and creating more efficient markets. *Huffington Post*, November 9.

Neapolitan, R., and Jiang, X. (2012). *Contemporary Artificial Intelligence*. London: Chapman Hall/CRC.

Nilsson, N. (1998). *Artificial Intelligence: A New Synthesis*. Burlington, MA: Morgan Kaufman.

Olden, J. D., and Jackson, D. A. (2002a). A comparison of statistical approaches for modelling fish species distributions. *Freshwater Biology* 47(10): 1976–1995.

Olden, J. D., and Jackson, D. A. (2002b). Illuminating the "black box": A randomization approach for understanding variable contributions in artificial neural networks. *Ecological Modelling* 154(1–2): 135–150.

Olden, J. L., Lawler, J. J., and LeRoy Poff, N. (2008). Machine learning methods without tears: A primer for ecologists. *Quarterly Review of Ecology* 83(2): 171–193.

Pang, B., Lee, L., and Vaithyanathan, S. (2002). Thumbs up? Sentiment classification using machine learning techniques. In *Proceedings of the Conference on Empirical Methods in Natural Language Processing (EMNLP)*. Stroudsburg, PA: Association for Computational Linguistics, 79–86.

Pearl, J. (1985). Bayesian networks: A model of self-activated memory for evidential reasoning. UCLA Computer Science, Irvine, CA: Tech. Rep. CSD-850017.

Peterson, A. T., and Vieglais, D. A. (2001). Predicting species invasions using ecological niche modeling: New approaches from bioinformatics attack a pressing problem. *BioScience* 51(5): 363–371.

Phillips, S. J., Anderson, R. P., and Schapire, R. E. (2006). Maximum entropy modeling of species geographic distributions. *Ecological Modelling* 190(3–4): 231–259.

Prasad, R., Dinesh, N., Lee, A., Miltsakaki, E., Robaldo, L., Joshi, A., and Webber, B. (2008). The Penn Discourse Treebank 2.0. In *Proceedings of the 6th International Conference on Language Resources and Evaluation (LREC)*, Marrakesch, Morocco, June. Paris: European Language Resources Association.

Poole, D., Mackworth, A., and Goebel, R. (1998). *Computational Intelligence: A Logical Approach*. New York: Oxford University Press.

Porter, M. (1980). An algorithm for suffix stripping. *Program Automated Library and Information Systems* 14(3): 130–137.

RAGE Frameworks. (2012). An extensible, model-based multi-pass approach for coreference resolution. RAGE White Paper. Westwood, MA.

RAGE Frameworks. (2013). An unsupervised approach to semantic map and meaning resolution. RAGE White Paper. Westwood, MA.

RAGE Frameworks. (2014). On information diffusion in financial markets – Evidence from news, blogs and social media. RAGE White Paper. Westwood, MA.

Raghunathan, K., Lee, H., Rangarajan, S., Chambers, N., Surdeanu, M., Jurafsky, D., and Manning, C. (2010). A multi-pass sieve for coreference resolution. In *Proceedings of the 2010 Conference on Empirical Methods in Natural Language Processing*, MIT, Cambridge, MA, October. Stroudsburg, PA: Association for Computational Linguistics, 492–501.

Reitter, D. (2003). Simple signals for complex rhetorics: On rhetorical analysis with rich-feature support vector models. *LDV Forum* 18(1–2): 38–52.

Rich, E. (1983) *Artificial Intelligence*. New York: McGraw-Hill.

Ripley B. D. (1996). *Pattern Recognition and Neural Networks*. Cambridge, UK: Cambridge University Press.

Rousseau, F., Kiagias, E., and Vazirgiannis, M. (2015). Text categorization as a graph classification problem. In *Proceedings of the 53rd Annual Meeting of the Association for Computational Linguistics and the 7th International Joint Conference on Natural Language Processing*, Beijing, China, July 26–31. Stroudsburg, PA: Association for Computational Linguistics, 1702–1712.

Russell, S. J., and Norvig, P. (2003). *Artificial Intelligence: A Modern Approach*, 2nd ed. Upper Saddle River, NJ: Prentice Hall.

Sagae, K. (2009). Analysis of discourse structure with syntactic dependencies and data-driven shift-reduce parsing. In *Proceedings of the 11th International Conference on Parsing Technologies*. Stroudsburg, PA: Association for Computational Linguistics, 81–84.

Salski, A., and Sperlbaum, C. 1991. A fuzzy logic approach to modeling in ecosystem research. In *Uncertainty in Knowledge Bases, 3rd International Conference on Information Processing and Management of Uncertainty in Knowledge-Based Systems, IPMU '90*, Paris, France, July 2–6, 1990; *Lecture Notes in Computer Science*, Vol. 521, ed. by B. Bouchon-Meunier et al. Berlin : Springer, 520–527.

Schick, A. G., Gorden, L. A., and Haka, S. (1990). Information overload: A temporal approach. *Accounting Organizations and Society* 15: 199–220.

Schneider, S. C. (1987). Information overload: Causes and consequences. *Human Systems Management* 7: 143–153.

Schroder, H. M., Driver, M. J., and Streufert, S. (1967). Human information processing—Individuals and groups functioning in complex social situations. New York: Holt, Rinehart, Winston.

Shapira, R., Y. Freund, P. Bartlett and W. Lee. (1998). Boosting the margin: A new explanation for the effectiveness of voting methods. *Annals of Statistics* 26 (5): 1651–1686.

Simnet, R. (1996). The effect of information selection, information processing and task complexity on predictive accuracy of auditors. *Accounting, Organizations and Society* 21: 699–719.

Simpson, C. W., and Prusak, L. (1995). Troubles with information overload—Moving from quantity to quality in information provision. *International Journal of Information Management* 15: 413–425.

Singh, P. (2002). The open mind common sense project. Available: http://www.kurzweilai.net/.

Soricut, R., and Marcu, D. (2003). Sentence level discourse parsing using syntactic and lexical informa- tion. In *Proceedings of the 2003 Conference of the North American Chapter of the Association for Computational Linguistics on Human Language Technology*. Stroudsburg, PA: Association for Computational Linguistics, 149–156.

Sowa, J. (1987). Semantic networks. In *Encyclopedia of Artificial Intelligence*, S. Shapiro, ed. New York: Wiley.

Sparrow, P. R. (1998). Information overload. In The *Experience of Managing: A Skills Workbook*, eds. K. Legge, C. Clegg, and S. Walsh, pp. 111–118. London: Macmillan.

Stockwell D. R. B., Noble I. R. (1992). Induction of sets of rules from animal distribution data: a robust and informative method of analysis. *Mathematics and Computers in Simulation* 33(5–6): 385–390.

Subba, R., and Di Eugenio, B. (2009). An effective discourse parser that uses rich linguistic information. In *Proceedings of Human Language Technologies: The 2009 Annual Conference of the North American Chapter of the Association for Computational Linguistics*. Amsterdam: Elsevier, 566–574.

Stoyanov, V., Gilbert, N., Cardie, C., and Riloff, E. (2010). Conundrums in noun phrase coreference resolution: Making sense of the state-of-the-art. In *Joint Conference of the 47th Annual Meeting of the ACL and the 4th International Joint Conference, Suntec*, Singapore. Stroudsburg, PA: Association for Computational Linguistics, 656–664.

Sutton, C. D. 2005. Classification and regression trees, bagging, and boosting. In *Handbook of Statistics: Data Mining and Data Visualization*, C. R. Rao et al., eds. Amsterdam: Elsevier, 24: 303–329

Turing, A. (1950). Computing machinery and intelligence. *Mind LIX* (236): 433–460.

Whitelaw, C., Garg, N., and Argamon, S. (2005). Using appraisal groups for sentiment analysis. In *Proceedings of the ACM SIGIR Conference on Information and Knowledge Management (CIKM)*. New York: ACM, 625–631.

Wiebe, J. (2000). Learning subjective adjectives from corpora. In *Proceedings 17th National Conference on Artificial Intelligence and 12th Conference on Innovative Applications of Artificial Intelligence*. Cambridge: MIT Press/AAAI Press, 735–740.

Winston, P. H. (1984). Artificial Intelligence. Reading, MA: Addison-Wesley.

Wurman, R. S. (2001). *Information Anxiety 2*. London: Macmillan.

Vega, C. (2006). Stock price reaction to public and private information. *Journal of Financial Economics* 82: 103–133.

Vollmann, T. E. (1991). Cutting the Gordian knot of misguided performance measurement. *Industrial Management and Data Systems* 1: 24–26.

Xu, F. (2007). Bootstrapping relation extraction from semantic seeds. PhD thesis. Saarland University.

CHAPTER 4

THE INTELLIGENT ENTERPRISE OF TOMORROW

4.1 THE ROAD TO AN INTELLIGENT ENTERPRISE

At the core of all enterprises is their commercial or business architecture. How are they set up to function? How are they designed to operate to maximize value to their stakeholders? How are they set up to compete? How are they designed to ensure responsiveness to the market? These are core questions that an enterprise architecture should revolve around as shown in Figure 4.1.

Post Adam Smith, enterprise architecture was based on the principle division of labor and centered on specialized functions. Each function was supposed to become highly competent with the dedicated focus and skill of the functional teams and to prevent skill fragmentation in producing the end product at scale. This was an environment where business technology was nonexistent. So, the only way to gain scale was to specialize. Such a view did not fully account for the silos that it would create in an organization and did not have the customer interests fully in sight. As a result there have been many proponents of a business-process centric architecture. Hammer and Champy (1993, 2004) forcefully articulated this position in their groundbreaking work. In the modern era with an increasingly flat world and with a significantly different technology landscape, a process-centric architecture has gained much momentum.

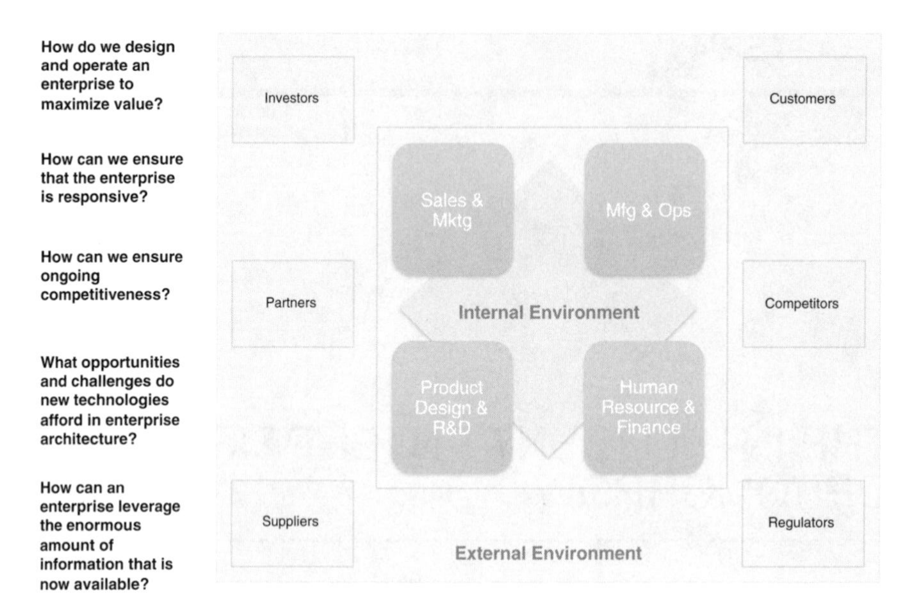

Figure 4.1 Enterprise environment

Figure 4.2 illustrates a typical functional architecture. Many corporations are still organized on a functional, hierarchical in their organizational divisions. Functional organizations with few exceptions often produce a less than optimal experiences for customers. There are frequent breakdowns in the processes' executions, resulting in a scramble to protect their brand in reaction to a customer's bad experience.

On the one hand, the advantages of a functional organization are that it is easier to adhere to standards because different groups specialize in a function or task, it expedites function-specific or vertical information flow, achieves

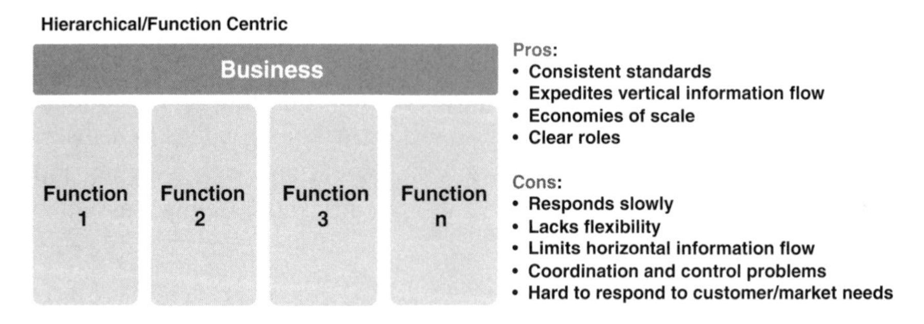

Figure 4.2 Functional enterprise architecture

scale economies because each functional group is dedicated to just that function, and there is clarity in roles. On the other hand, a functional organization is typically much slower to respond to an external need because it has to coordinate action or response across multiple functions, lacks flexibility at a process level, and does not provide adequate visibility or information flow across a business process that spans multiple functions. A functional organization has been found to lead to significant coordination and control issues and to be unresponsive to customer or market needs.

A process-centric organization, in contrast, revolves around end to end business processes. While a function-centric organization has enabled enterprises to increase scale in the post–Industrial Revolution era, dramatic increases in scale, along with specialization, have led to organizational silos and have made them less responsive to the market, changes in customer expectations fueled by the Internet, and mobile communications. This weakness of the function-centric organization has become more apparent in today's highly dynamic world in which it is a critical competitive necessity for businesses to adapt and innovate their processes end to end.

Figure 4.2 depicts a process-centric organization with business processes consisting of several functions. The processes may be interconnected at higher levels of the enterprise. Business processes are typically part of a bigger picture. Business owners have the responsibility to maintain optimal business processes at all times. They have control over the functional resources required in the course of executing any one of their business processes.

The advantages of a process-centric organization are many. As listed in Figure 4.3, process-centricity eliminates functional silos and improves the organization's ability to respond to customers or market needs. It also promotes innovation at a business process level instead of just at a task level.

In reality, few organizations have been able to achieve true process re-orientation, largely because of the limitations of technology and the significant culture change involved in transforming their orientation. Nevertheless, we find that often the accumulated frustrations from the inability of a function-centric organization to respond to market needs eventually lead to major organizational transformation projects, and thus essentially force a transformation to process-orientation. The slogans of such projects often reflect these frustrations – "One and done," "Global one," and so on.

There is another dimension in enterprise design that is worth mentioning. This is the difference between design and execution in the enterprise architecture (Figure 4.4). Design refers to the design of strategies, business models, and business processes. However, once the strategy is set, the business processes are designed to be executed repeatedly by the enterprise's production

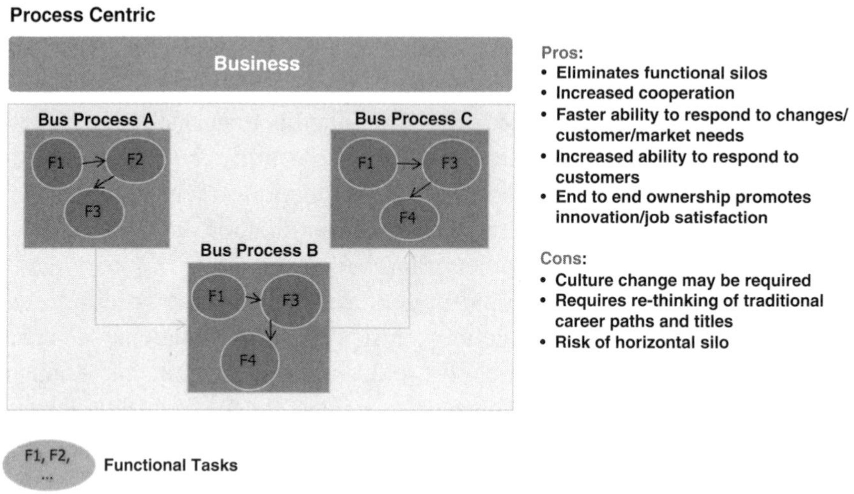

Process Centric

Business

Bus Process A: F1 → F2 → F3

Bus Process B: F1 → F3 → F4

Bus Process C: F1 → F3 → F4

F1, F2, ... Functional Tasks

Pros:
- Eliminates functional silos
- Increased cooperation
- Faster ability to respond to changes/ customer/market needs
- Increased ability to respond to customers
- End to end ownership promotes innovation/job satisfaction

Cons:
- Culture change may be required
- Requires re-thinking of traditional career paths and titles
- Risk of horizontal silo

Figure 4.3 Process-centric enterprise architecture

team. Often in these enterprises some members of the production team will assist in designing the enterprise architecture. Regardless of whether an organization is function or process centric, only a few members will be involved in the design of the architecture; the bulk of the organization's members will be involved in execution.

Every organization attempts to create a continuous feedback loop between the execution and design parts of their business processes. Where this is well

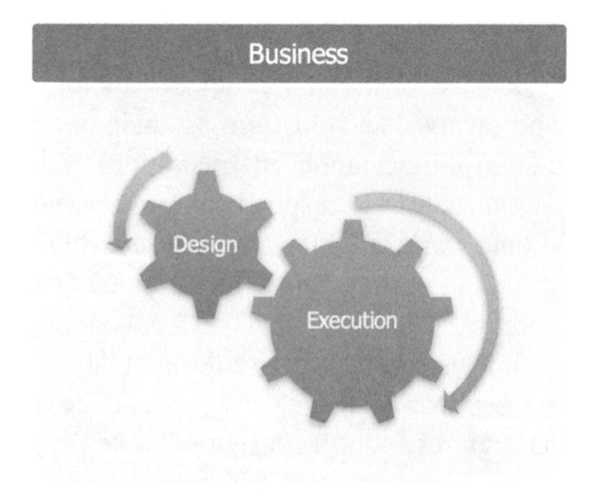

Figure 4.4 Design versus execution in enterprise architecture

coordinated, all works very well. In most circumstances, there is no systematic design. Issues build up over time; when things get to a point where there are very noticeable, there is a reaction, and the process improvements get attention. This is not because enterprises lack the desire to effect continuous improvements but rather because of the huge effort involved usually in effecting even small changes to processes due to inflexible or nonresponsive technology.

4.2 ENTERPRISE ARCHITECTURE EVOLUTION

As we noted in Chapters 2 and 3, the recent advances in technology offer exciting prospects of change in enterprise architectures. Our thesis is that these advances will enable an enterprise to reach what we are calling an "intelligent" state. Moreover, this future architecture will have major implications for the role of humans in the workplace and the type of work humans do. Enterprises are already deploying intelligent machines that automate much of what humans do on the execution side, and in time, machines will likely extract insight for humans with respect to the design of various business processes. In this chapter, we attempt to paint a road to such an intelligent enterprise.

Before we lay out an architecture for the enterprise of tomorrow, we will briefly trace the evolution of such architectures in conjunction with technology waves and the structure of work over time.

4.2.1 Technology Evolution

In our sketch of a roadmap for the intelligent enterprise, there is a prominent influence of business technology. Technology has evolved, leaving an indelible mark on all aspects of business processes (Figure 4.5). It has automated mundane and mechanical work, and staying up to date with technological advances has become a major source of competitive strength for any business.

Technology 1.0 Every major stage in the evolution of technology has had a corresponding impact on the evolution of the enterprise's execution architecture. As technology paradigms have evolved, enterprise architectures have adapted and changed, leveraging the new possibilities and challenges. Each successive stage of technology evolution has in turn triggered a cascade of

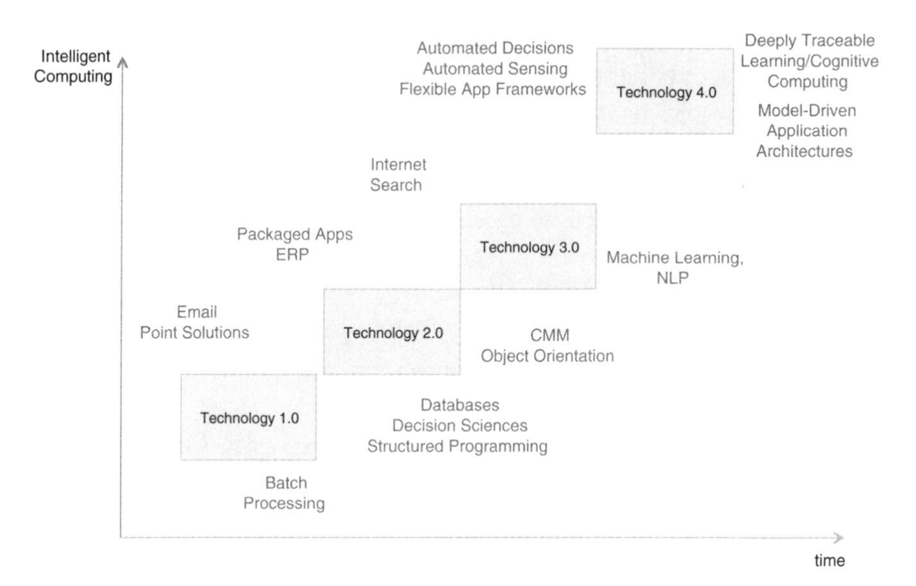

Figure 4.5 Technology evolution

changes in how enterprises design themselves. More recent advances in technologies demand more dramatic changes in how enterprises might rationalize their functions today and in the future.

In the beginning, computers were mostly used for recordkeeping in maintaining payroll and financial records. I recall it being a big deal in the late 1970s to close financial books by the 10th day of the following month. All transactions were recorded on paper and then processed at the end of the month. Using computers for decisions and analytics was a distant dream. Yet in 30 to 40 years since, we have seen quantum leaps in computing capabilities.

About two to three decades ago, human–computer interaction began to move beyond recordkeeping to basic productivity and communication. A major breakthrough was email. And recordkeeping on mainframes started to appear in smaller desktop computers, called mini-computers. Much of the innovation in those years had to do with computer hardware and computing power. A lot of work was being driven by expensive data storage. From a business perspective, the 1960s IBM typewriter gave way to the 1970s word processing machines like the Wang, and then in the late 1980s quickly to word processors on PCs.

Technology 2.0 Then came the personal computer. The introduction of the IBM PC XT and IBM PC AT in 1984 and 1985 came at about the time software applications emerged to improve basic productivity (word processing, spreadsheet, presentation), and with this development, point solutions for more involved business processes started to emerge. Simultaneously, a major breakthrough came about in computing with the separation of data from computer programs and the birth of dedicated databases. Codd and Date (1982) introduced the relational data model, an axiomatic approach to data representation that provided an assured level of data integrity as we started to manipulate data in databases.

As point solutions started to proliferate, businesses realized that these applications were inflexible, and expensive to upgrade and enhance. There was discussion that a better way may be in the form of standardized applications at least for non-core business processes like resource planning and financial accounting. ERP and packaged solutions emerged. Structured programming was introduced in the 1980s. Client–server architectures evolved to shift logic down from mainframes and leverage the capabilities of the desktop.

Methodologically, Carnegie Mellon University's Capability Maturity Model (CMM) enabled software development to evolve from an art form with a high level of reliability and quality issues to more of a science. The focus on methodological approaches would evolve the software development lifecycle (SDLC) into an engineering science and increase the predictability and reliability of software development in general. Nevertheless, these approaches were not by and large able to address time to market, flexibility, and knowledge/insight challenges. The evolution in SDLC attempted to make the SDLC more efficient, but it failed to dramatically change the SDLC itself.

Technology 3.0 The most significant part of Technology 3.0 was the widespread connectivity with the public availability in the 1990s of the Internet. This heralded an unprecedented wave of change in the technology landscape. Browser-based applications became popular, going back to the server-centric model from the mainframe days. Machine learning quietly powered Internet search, and all of a sudden, management science and algorithmic models started to resurge. Natural language processing spilled out of the academic labs into the real world; computational linguistics got a real fillip with the development of key enabling utilities like Wordnet.

Many open source technology frameworks emerged in this period, such as Java, J2EE, Spring/Hibernate, .Net., Javascript, Eclipse, and a large number of Open Source tools. These have made programming easier and faster and afforded portability across operating systems. All were developments that were unimaginable during the Technology 2.0 phase.

Technology 4.0 Each technology phase has ushered in a new wave of capabilities and impact. However, we believe the next phase of technology evolution that is upon us will bring in an unprecedented amount of change in the way enterprises are designed and function.

As was discussed in Chapter 2, the goal of business technology in the context of an enterprise can be viewed through the four challenges that it needs to address – reliability, flexibility, time to market, and knowledge. Technology 3.0, while clearly a huge improvement over previous technology capabilities, did not solve fundamental issues of time to market, the need to cope with the rapid changes in business needs, and the dependence on the availability of specialized resources. The emergence of the Internet has meant that businesses now have to address an information explosion. Quite recently corporations have started to adopt Agile methodologies. As with anything new, we have yet to see what Agile methodologies will accomplish. An Agile methodology should, however, not be confused with improving time to market or even flexibility. The principal benefit of an Agile method is to improve reliability, to increase the probability that the end product will meet business needs. It is really important to realize that *Agile does not equal agility*. Adoption of agile methodologies will not solve the time to market and flexibility challenges.

Nevertheless, this current phase of technology evolution has introduced metadata-driven model-oriented flexible software platforms that make near real time software development possible, without any programming. It will no longer take a year or two to create an enterprise scale mission critical application. Software will be fundamentally process oriented from the ground up, allowing businesses complete flexibility in modifying software as needed, literally on the fly. This is likely to have a huge impact on the software programming industry. Dare we say, programming will be relatively extinct 10 to 20 years from now!

The other development that has significant implications for all enterprises is the maturation of AI technologies. AI is rapidly emerging out of R&D labs into the mainstream. Computational statistics have achieved significant success in computer vision applications – such as with autonomous vehicles and facial recognition. Deep learning technologies are now being successfully applied in mainstream business processes as illustrated in several chapters in this book. While computational statistics based machine learning will be successful in some Domains like computer vision, traceable machine learning methods will gain popularity and trust. Specifically, we believe that traceable approaches to understand natural language will have enormous impact on building intelligent machines. The 2010s promise to be the age of machine intelligence. Among other things, this development has significant implications for low-cost, manual outsourcing. Intelligent machines will likely automate knowledge-based work.

As mentioned, each stage of technology evolution has had a definitive impact on enterprise design and architecture. Let us now trace our view of this evolution.

Enterprise 1.0 – People Led, Manual; Intractable Information Enabled by Technology 1.0, enterprise architecture in the 1960s was at this stage predominantly centered on people. Employees knew how business processes were to be executed; process knowledge was only in the minds of employees, and they acted on information. Information flowed from person to person in a manual fashion, and employees performed their functions. This stage of enterprise architecture was enabled substantially by email and mainframe computing. Both design and execution was done entirely manually.

In this phase, as illustrated in Figure 4.6, businesses were completely dependent on people's memories and capabilities, and there was little or no institutionalization of business processes. The more organized or larger businesses had volumes of procedure manuals on paper. Each person would complete his or her task in a process and approach the next person to do the next task. When in doubt, employees could consult their procedure manuals. Errors were quite common. Scaling and business innovation were much more constrained.

Enterprise 2.0 – Islands of Inflexible Automation, Rigid Information Flows Enabled by Technology 2.0, enterprise architecture began to reflect the influence of business technology (Figure 4.7). Businesses began to implement

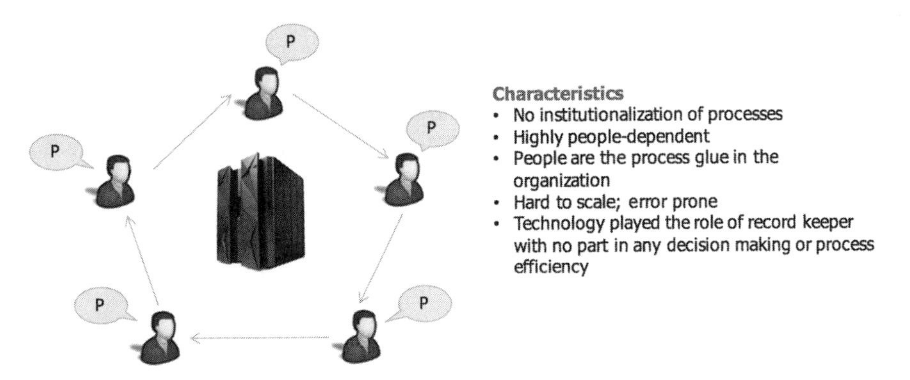

Figure 4.6 Enterprise 1.0

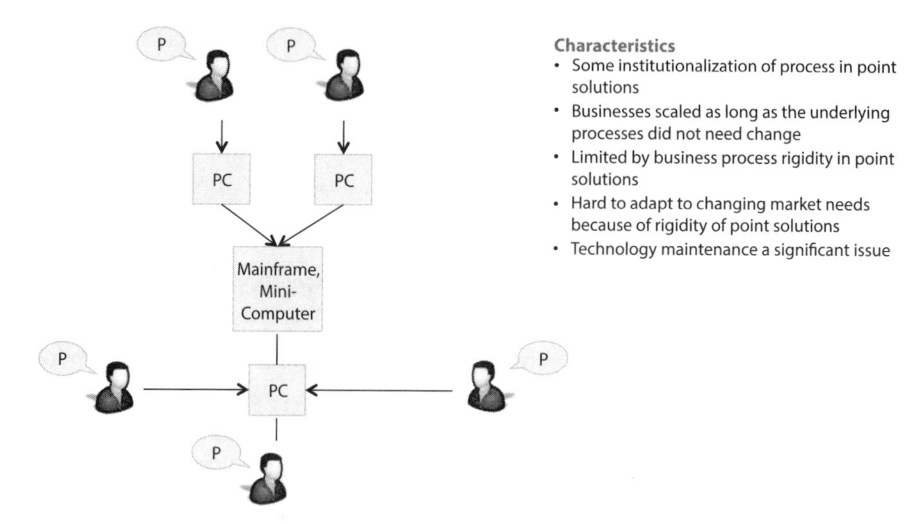

Figure 4.7 Enterprise 2.0

point solutions and created islands of automation. The discovery of the ability to use technology in the form of software applications lead to an explosion of home-grown, custom applications. The flow of work and information was somewhat automated, but the actual work was still left to the people.

For sure, there were point solutions and technologies that enabled people to do their work far more effectively compared to Technology 1.0. Business processes were hard-coded in these point solutions. As long as business processes did not change and the applications were robust enough, businesses could scale. But any change would be crippling and slip them back to Enterprise 1.0.

Software development was still largely an art form in the 1980s. Over time, driven by the high maintenance costs of custom applications and the failure of most software development projects, modular, structured programming were introduced and later programming standards, and eventually the CMM made software development more of a science. This phase was characterized by high rates of failures of software development projects and a high level of unpredictability in the realization of the promise of business technology.

Software performed point acts of automation. Flexibility was lacking. While there was general realization that enterprises had to be process driven, most of it was the lexicon of the strategy consultant and little could be implemented in practice. Technology was lagging behind the need for flexibility and change.

Enterprise 3.0 – Standard Packaged Applications; Workflow Stung by the failures of most software development efforts and high maintenance of custom applications, enterprises moved to adopt standardized packaged software products. This, of course, meant that enterprises were substantially relying on somebody else's interpretation of business processes in their industry and hoped they could configure it to suit their needs. Enterprise resource planning (ERP) systems came onto the scene. ERP implementation became its own industry in the 1990s. Many consulting companies set up huge practices just implementing ERS applications like SAP and Oracle.

When you read public reports of these implementations, it will be obvious that they were rigid, inflexible, and expensive. In fact, in the 1990s, there were so many horror stories on how implementations went terribly wrong or resulted in huge cost overruns in large multinational corporations. I know of so many Fortune 500 companies where people were scratching their heads trying to work around the long-phased implementation of an ERP application.

During these implementations, often different divisions or business units would end up being on different ERP applications and not talk to each other. This left the operational teams to cobble together some patchwork solution or resort to manual work to get their work done. For example, credit managers would have to look up multiple systems and add up exposures manually to get total exposure for a customer. Or accounts receivable personnel would not be able to look across divisions to see what the customer owed, and so on.

The pervasive adoption and use of the Internet acted as a catalyst for a new class of packaged applications that emerged in this period. These applications modified the idea of owning a license to the application to the idea of using the application as a service. Today, SaaS or cloud-based applications like Salesforce have become very popular. Salesforce has been a runaway success largely because of ease of adoption, a disruptive economic model, and a built-in development environment. However, we see the same level of process rigidity in SaaS applications. "Cheaper" does not solve the fundamental issues of time to market and flexibility.

These packaged products have at best been able to provide a few knobs and levers to change some specific functionality in the product. Their implementation of the end to end processes underlying these applications are largely fixed. In Figure 4.8, we depict this by the process inside ERP and CRM applications – the three connected Ps. This process is fixed inside these applications. They can be modified by custom programming. But the fixed Ps create rigidity. With these packaged products, enterprises are still far from being able to adapt rapidly to changes in market conditions. Most business enterprises are still in a continual state of inefficiency but have gotten accustomed

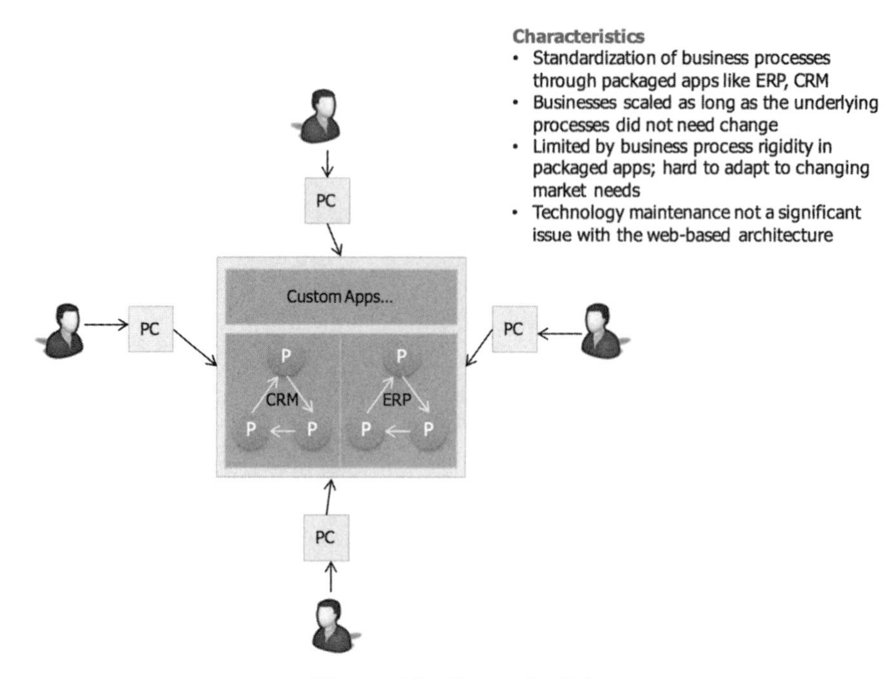

Figure 4.8 Enterprise 3.0

to long technology build cycles and their relative inability to adapt rapidly to market-driven changes.

Enterprise 4.0 – The Intelligent Enterprise of Tomorrow Technology 4.0 has the potential to cause dramatic changes in enterprise architecture. This is obviously the reason we think the transition to Enterprise 4.0 will imply a much greater amount of change than the previous waves of change. Figure 4.9 shows Enterprise 4.0 at a significantly higher altitude than Enterprise 3.0. The enterprise of tomorrow has the ability to move to an entirely different plateau. The last 5 to 10 years have seen enormous advances in technology capabilities that make it feasible to have a very different enterprise architecture. Such architecture will enable the realization of the "intelligent enterprise" with very different roles for humans and machines. The "future of work" will be dramatically different.

As we have explained in Chapters 2 and 3, we see three key drivers of this shift to a different plateau – flexible, near real time software development; the ability to create intelligent machines that will do knowledge-based work; and the availability of enormous information. We briefly recap these three drivers and introduce an architecture that reflects these drivers in Enterprise 4.0.

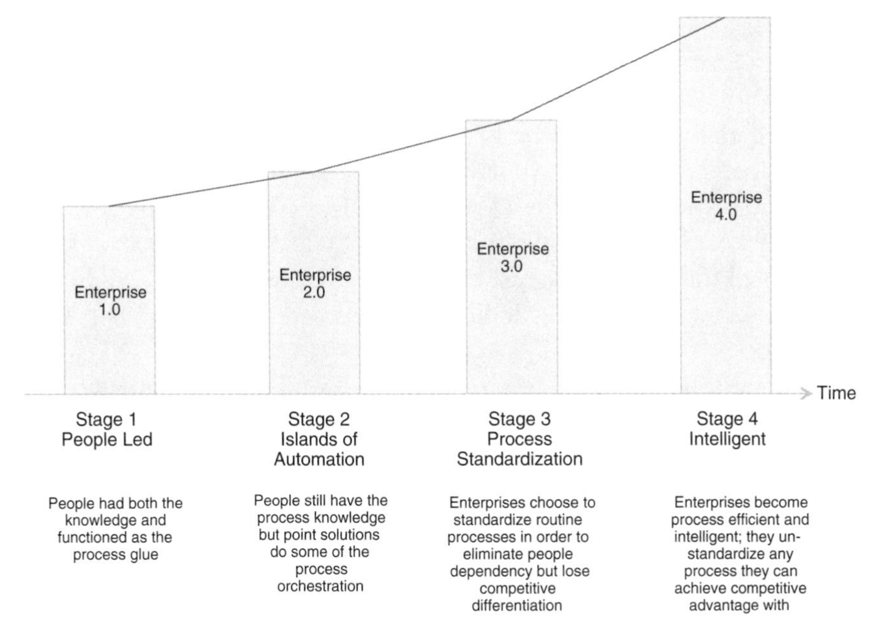

Figure 4.9 Intelligent enterprise of tomorrow

4.2.2 Flexible, Near Real Time Software Development

As described in Chapter 2, meta model-driven technology frameworks will enable real time software development without any or little programming. These frameworks are business process oriented from the ground up. They provide the ability to configure the entire business process to the needs of a specific enterprise. They are designed to do away with programming entirely but are extensible to include programming as needed. They rely on an entirely metadata-driven architecture using an abstract business process as the "model." The implications are that entire applications and all their business processes will be reduced to data and stored as data.

Imagine an application as a set of flow charts that are its business processes; imagine being able to implement these flow charts visually using building blocks from the technology framework. These blocks are abstract and context agnostic. The modeler or business analyst provides the context, as metadata, to such abstract blocks. The runtime engine of the framework can execute the business process, including dynamic user interface, at runtime by caching all the metadata for the business process model.

Such a framework will truly enable enterprises to design and implement business processes specific to their businesses and environments in a near real time fashion. These frameworks will completely eliminate application-specific programming over time.

On top of such horizontal technology frameworks, vertical solution frameworks will emerge to reflect best practices in different Domains. These are pre-packaged process frameworks that are exemplary models of the business processes needed for the Domain. The LiveWealth™, LiveCredit™, and LiveSpread™ frameworks from RAGE Frameworks are examples of such vertical frameworks. They are in use at global enterprises and provide them with rapid, unlimited flexibility as well as with significant competitive advantages.

These vertical frameworks can be contrasted with packaged applications. The vertical frameworks allow enterprises to configure the frameworks from the ground up with very little effort and no specialized skills like programming. The ability to let enterprises configure their frameworks without programming will unleash a lot of innovation in the packaged application industry in the years to come. We envision a future where all packaged applications will be completely configurable from the ground up. Similarly, we expect a lot of disruption in the software programming industry. Platforms like RAGE AI™ turn the conventional forward-oriented software development lifecycle upside down, as discussed in Chapter 2. We envision a significantly reduced need for programming in general.

4.2.3 Machine Intelligence

The second key driver for Enterprise 4.0 setting it on an elevated rate of change is machine intelligence. As explained in Chapter 3, the availability of huge amounts of information on a real time basis has resulted in both an information overload and the opportunity to derive insights previously deemed impossible. This has provided the impetus for the maturation of natural language technologies and for a re-focus on the "analytics of such an enormous quantum of information," also referred to popularly as BigData Analytics. Such analytics are envisioned to be capable of analyzing structured data and/or natural language. Structured data analytics is not necessarily new, albeit the volume and the algorithmic approach to it are relatively new. Analysis of natural language text is quite new. We refer to machine intelligence as the automation of such analytics – from acquiring knowledge, to automatically drawing inferences from the vast amount of data, to automated integration of such inferences with the rest of the business processes.

The rapid maturation of BigData Analytics, particularly the ability to analyze unstructured text, offers an exciting opportunity for enterprises to acquire intelligent machines that are dedicated to different business processes. Machine intelligence can be created even for fuzzy processes like competitive intelligence in a systematic and meaningful way. Successfully leveraging this opportunity is at the heart of the intelligent enterprise of tomorrow and will be a competitive necessity.

4.2.4 E4.0 Architecture

As we have already discussed this at length in previous chapters, the rate of information availability and growth is only accelerating. Suffice it to say that with every device now beginning to emit information, we ain't seen nothing yet!

This brings us to Enterprise 4.0. In E4.0, we envisage a big change in the way humans interact with the machines. Intelligent machines will have a far greater part in the day to day functioning of the enterprise. We imagine such an architecture (Figure 4.10) to have intelligent machines dedicated for specific business processes and domains.

Such machines will be built on a platform like RAGE AI™ and will perform two functions: provide insight to humans to enable the optimal design of intelligent business processes through continuous analysis of vast amounts

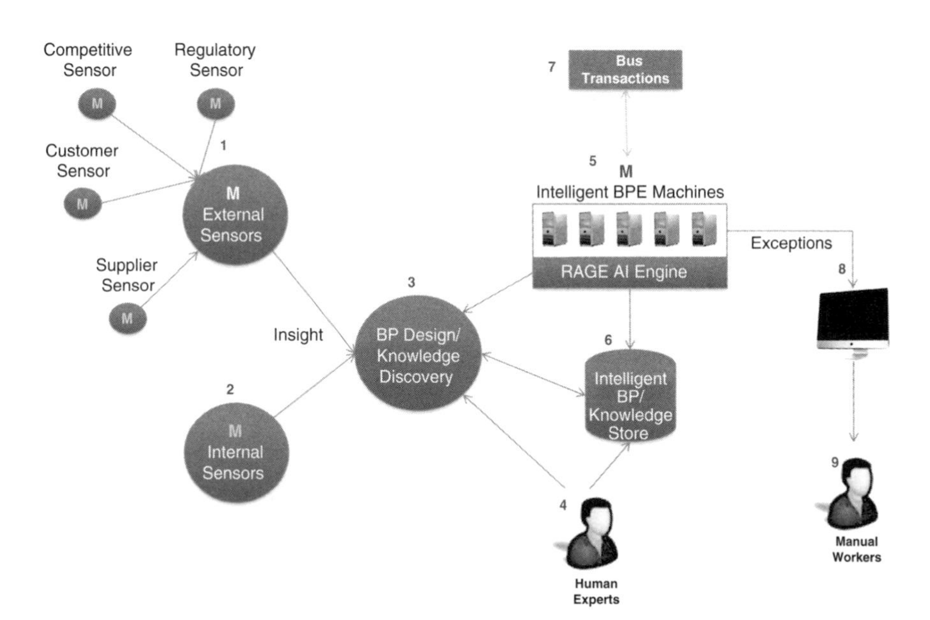

Figure 4.10 Intelligent enterprise of tomorrow

of data, and automatically execute business transactions without human assistance based on an automated business process designed by humans. Analytical machines will acquire their knowledge from continuous analyses of vast amounts of data. Such knowledge and insight will enable humans to refine their designs.

Humans will perform three important roles: design intelligent business processes, refine the automated knowledge machines acquire, and process exceptions that could not be handled by the machine. Intelligent business processes will be designed and implemented on a near real time basis using RAGE AI™. Since automated knowledge acquisition by machines will become reliable and complete only over time, humans will be needed to process exception transactions. Humans will further be needed to curate the knowledge gathered by machines, or seed machines with initial knowledge, or to provide more extensive knowledge where machines are unable to gather it.

The intelligent enterprise architecture in Figure 4.10 consists of nine components:

1. *External sensors* These are intelligent machines that are continuously sensing the external environment from various aspects. For example, a competitive sensor can scan the external environment to monitor developments with competitors, new innovations, business models, and the like. These sensors are seeded with knowledge provided by Human experts identified as item 4. Such knowledge is contained in the Knowledge store identified as item 6. It can take the form of a skeletal concept map as described in Chapter 3. The RAGE AI™ platform has the ability to auto-discover knowledge on top of the seed knowledge provided by the Experts. Over time, as these intelligent machines gain ever more knowledge, they will be able to provide insights that will help the Human experts improve their products and processes as illustrated by item 3 – BP design.
2. *Internal sensors* Similar to external sensors, internal sensors monitor and analyze internal knowledge contained in appropriate documents. Many firms have a depth of knowledge and information in internal documents that is generally not leveraged adequately. The internal sensing machines will also be seeded by the Human experts with an appropriate concept map.
3. *BP design/Knowledge discovery* This component is the actual design of intelligent business processes and/or performs automated knowledge discovery using RAGE AI™. BP design can be done by the experts.

Knowledge discovery can be automatically done by the machines optionally based on seed concept maps provided by the experts.

4. *Human experts* Experts design business processes for the execution environment. Their design will reflect the business models and objectives of the business processes. Such design will be informed by any insight that the sensors discover on an ongoing basis. Experts also can seed the knowledge discovery process by providing their knowledge on the business and various dimensions of the business such as the industry and relevant products.
5. *Intelligent BPE machines* This is the primary execution environment. For reference, this environment comprises packaged and custom applications in today's context. In Enterprise 4.0, such an environment will comprise intelligent machines that are driven by RAGE AI™ and execute business processes to handle business transactions. As many transactions as the BP design permits will be automatically handled. Exceptions will be directed to humans through an exceptions process as iillustrated with items 8 and 9.
6. *Intelligent BP/Knowledge store* This store will contain all the business processes designed by the experts and also all the knowledge accumulated by the machines. Technically, such a store can take many forms, like a database, file system and so on. As discussed in detail in Chapter 2, these business process models will use abstract components in a meta model-driven platform like RAGE AI™ and will eliminate programming.
7. *Business transactions* These are the business transactions the enterprise engages in. They will originate from outside the enterprise. In Enterprise 4.0, they will be handled by intelligent machines denoted by item 5.
8. *Exceptions* Transactions that cannot be decisioned or acted on by an intelligent machine will be flagged as exceptions requiring human intervention.
9. *Manual workers* Exceptions will be adjudicated by human workers. In Enterprise 4.0, the goal is to minimize or eliminate such exceptions. Periodically, the Human experts will analyze exceptions to identify improvements to their BP design to eliminate them in the future.

Implied in the foregoing architecture is a dramatic change in the structure and flow of work in an enterprise. To the extent a business transaction is processed automatically by an intelligent machine, there will be no role for humans in that process. Even when there is an exception, machines will drive the flow and allocation of work to humans for exception processing based on intelligence about the work and the workforce.

In E4.0, business processes can be designed and modified at will. Implementation will be near real time. Enterprises will no longer be shackled by the constraints of time with respect to software development. We expect this will embolden enterprises to rethink the adoption of inflexible packaged applications. It will also spur the packaged application industry to use platforms like RAGE AI^{TM} and fundamentally redesign their applications to be flexible at the business process level.

4.3 HUMANS VERSUS MACHINES

Does this mean that machines will replace humans? And that a large number of jobs will be eliminated? The idea that technology will obsolete jobs has been argued for centuries. Many pundits have routinely raised alarms that automation through new technology will results in massive numbers of job losses. We can all obviously see that despite significant technological advances over the last couple of centuries jobs have not gone away. As Akst (2013) points out, in the 1930s no less than John Maynard Keynes, the most influential economist of the twentieth century, fretted about temporary "technological unemployment," which he feared would grow faster than the number of jobs created by new technologies.

The anxieties surrounding the impact of technological automation on employment have resurfaced with a popular book, *The Second Machine Age*, in which Brynjolfsson and McAfee (2014) argue that technological progress is going to leave behind some people, perhaps even a lot of people, as it races ahead. In their opinion, there's never been a worse time to be a worker with ordinary skills and abilities to offer because computers, robots, and other digital technologies are acquiring such manual skills and abilities at an extraordinary rate.

We think these fears are quite realistic. What nobody knows is the rate of substitution of old jobs by new jobs as a result of the new technology automation wave. While there has already been a clear skill shift in jobs, the idea that technological automation will eliminate one job but create a job for a different skill or another person has been well recognized. Autor (2013) argues that trade (imports from China and elsewhere) has increased unemployment in the United States, while technology has reshaped the job market into something like an hourglass, with more jobs in fields such as finance and food service and fewer in between. There is widespread agreement that automation thus

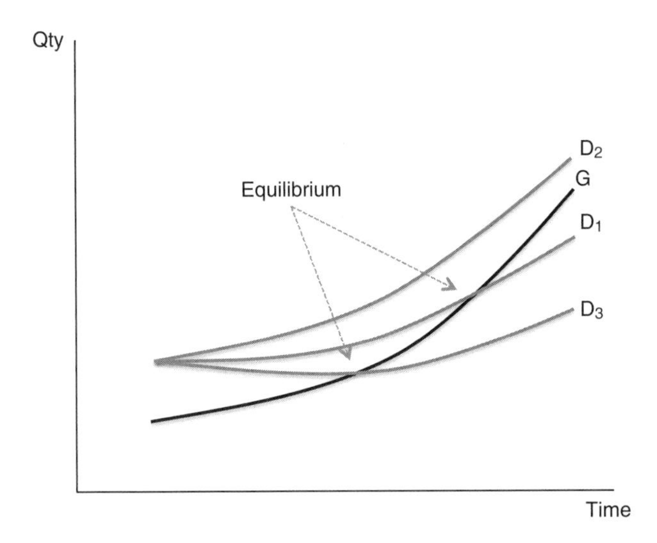

Figure 4.11 Rate of displacement versus rate of new skilled jobs

far has eliminated less-skilled jobs disproportionately and has created demand for more-skilled workers. Autor (2015) observes that in the last few decades, there has been a "polarization" of the labor market, in which wage gains have gone disproportionately to those at the top and at the bottom of the income and skill distribution, not to those in the middle.

Figure 4.11 illustrates the trade-off between technology-driven displacement and the creation of new jobs. Historically, the rate of new jobs has generally kept pace with the rate of displacement. However, the current resurgence of anxiety is the result of the same phenomenon that resurfaces with every major technology wave – we don't know the specifics of the interaction between the elimination of old jobs and the creation of new jobs. D_1, D_2, and D_3 are three hypothetical job displacement trajectories through automation. G is the hypothetical trajectory of new job creation requiring new skills. Ignoring the training time required for new jobs, there will be equilibrium at the point where the displacement will equal new jobs. There is a scenario where new jobs never catch up with displacement and this will result in a net loss of jobs. This is shown by D_2 and G.

It is important to recognize that these comparisons are at a macro level. When we look at specific industries and geographies, the impact may be quite different. For example, in the business process outsourcing (BPO) industry, where many of the jobs could be viewed as low skilled, the rate of displacement could be potentially disproportionately high.

We believe that this wave of automation will displace a lot of jobs in the BPO industry and will likely displace jobs to some extent in the knowledge process outsourcing (KPO) industry. It is imperative for BPO firms to start to rapidly transform themselves and leapfrog this technology wave.

I have always believed that the BPO market is not sustainable in the long run. In fact, in an interview to Knowledge@Wharton, I explain my rationale in some level of detail (Srinivasan, 2009). I am sure many in the BPO industry did see this wave coming, but no one felt under any pressure to change course until very recently. Now customers have started to demand technology-based leverage or "automation," and this finally is causing some BPO firms to recognize this challenge to their business model.

While recognizing the phenomenon of skill-based displacement, Autor (2015) argues that the tasks that have proved most vexing to automate are those that demand flexibility, judgment, and common sense. He concludes that it is unlikely that automation will extend to these Domains.

It is argued that machine learning technologies based purely on pattern recognition, without understanding the full context, have yet to reach a level of maturity where machines can substitute humans in mission critical situations. While IBM's Watson computer famously triumphed in the trivia game of *Jeopardy* against champion human opponents, it also produced a spectacularly incorrect answer on the second day to the question, "Its largest airport was named for a World War II hero; its second largest, for a World War II battle," as Toronto, a city in Canada (Kanalley, 2011).

One should recognize that the underlying technologies—the software, hardware, and training data—are all improving rapidly (Andreopouos and Tsotsos, 2013). Some researchers expect that as computing power rises and training databases grow, the brute force machine learning approach will approach or exceed human capabilities. Others suspect that machine learning will only ever "get it right" on average, while missing many of the most important and informative exceptions. Ultimately, what makes an object a chair is that it is purpose-built for a human being to sit on. Machine learning algorithms can have fundamental problems with reasoning about "purposiveness" and intended uses, even given an arbitrarily large training database of images (Grabner, Gall, and Van Gool, 2011).

Autor (2015) reaches a very important conclusion: "Moreover, an irony of machine learning algorithms is that they also cannot "tell" programmers why they do what they do." This conclusion forms one of the principal underpinnings of Chapter 3. Much of the current machine learning technologies are getting carried away with brute force methods based on computational statistics. They will work well for certain classes of problems but not for many. More important, they do not replicate the functioning of human brain as the proponents argue. That is why they cannot tell why they do what they do.

As outlined in detail in Chapter 3, however, a different breed of machine learning technologies, traceable and contextually relevant machine learning technologies like RAGE AI™ are making significant progress. RAGE AI™ will be successful in reproducing human intelligence over time because it replicate the way humans think more closely. We referred to a market intelligence application using RAGE AI™ that senses important and strategic developments in 20 sectors on a real time basis. Its accuracy across each topic for each sector is consistently more than 90% week after week. In the sales prospecting application also referred to in Chapter 3, the application finds 250% more relevant items than a team of 20 people who do the search manually for sales leads. In a contract review reconciliation use case for a large Fortune 500 firms, such an approach enables the identification of overbilling from vendors. In all these applications, the reasoning of the solution is completely traceable and transparent. The machines can tell us precisely why they do what they do.

We believe that traceable machine learning and near real time software development will gain traction in the coming decade. Such machine learning technologies in fact are the only approach that will be successful. Black boxes, like the many in the market today, will only work where traceability is not important and where the underlying data is highly homogeneous. These technologies will rely on contextually relevant Knowledge stores just like we humans develop over time. This is a more realistic reflection of how human brains work. It is foolish to think that machine learning will succeed without deep Domain-specific knowledge where the cause–effect relationship is transparent and verifiable. We can expect that there will be significant developments in knowledge acquisition so that machines won't have to take as much time as we humans currently do to acquire the same knowledge.

In the not so distant future, I do see many current skills getting completely eliminated and generally, a high level of skills dis-location. Some of these will be knowledge-based skills and jobs. Both the BPO and KPO (knowledge process outsourcing) industries will be severely affected. I also see the software programming industry getting heavily affected by the emergence of near real time software development with little or no programming. While programming will not become extinct anytime soon, we will need a lot fewer programmers per capita within the next couple of decades. Programmers will have to retool themselves to be business analysts, business process designers, and architects.

4.4 SUMMARY

The architecture of the future for a business enterprise is driven by two important developments related to business technology: near real time software development and artificial intelligence. We amalgamate in this chapter the ideas presented in Chapters 2 and 3 and introduce an integrated framework for these ideas.

Many enterprises are struggling to make sense of all the developments in technology and the information explosion. They know they should do something but are still trying to get their arms around what all this means for them. Many have active programs to leverage analytical methods to gain insights from the vast amounts of data that they have accumulated from the "big data" groups within their enterprises. Like most new technology-driven waves, there is a lot of hype around such groups. We believe that most such groups will fall short of their promise.

In this chapter, we make the argument that enterprises should not just think about big data in isolation. They should think more holistically about an architecture for the future and leverage all their technological developments in perspective. We provide a conceptual framework for such an architecture. In doing so, we elevate the current conversation around big data and machine learning in two directions. First, we urge enterprises to think about traceable machine learning and to embed them in intelligent machines. Enterprises should not blindly adopt black boxes. Second, enterprises should start to leverage near real time software development. This might mean that they move away from packaged applications and develop or buy solution frameworks that are process centric from the ground up. Such frameworks will enable them to achieve rapid time to market and flexibility in adapting to market needs.

APPENDIX: A FIVE-STEP APPROACH TO AN INTELLIGENT ENTERPRISE

Most enterprises are at various stages between Enterprises 2.0 and 4.0 with respect of most of their business functions. Figure 4A.1 gives a simple commonsense approach to setting up a transition to Enterprise 4.0.

The first step is obviously to define priority areas. This will typically be done by the senior leadership in the company. The transition will require significant change in management style, and it is important to have the support of the senior leadership in the company.

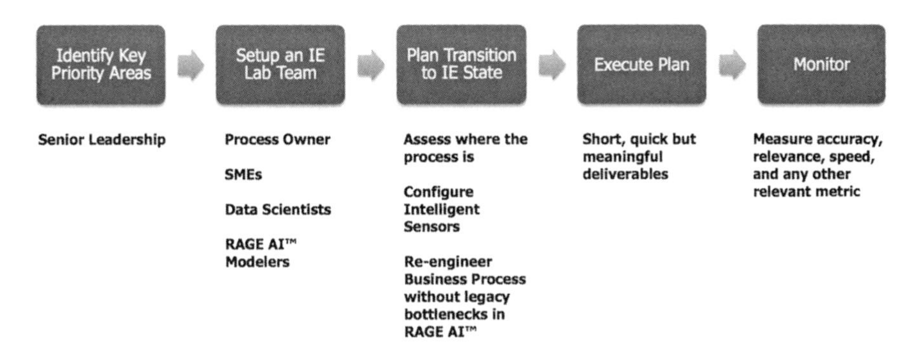

Figure 4A.1 Simple approach to an intelligent enterprise architecture

The next step is to set up an IE Lab for each of the priority areas. The IE Lab team is like a SWAT team. It needs to have all the relevant expertise for the respective business process. We envision such a lab team to have an owner, subject matter experts, and RAGE AI™ modelers.

The IE Lab team will need to assess the current state and map the future IE state and a transition plan. Such a plan might consist of phases that will get the business function to the IE state.

The execution step will be characterized by rapid deliveries. Each delivery of a meaningful unit of change should typically not take more than 15 to 30 days. This is entirely possible with technology platforms like RAGE AI™.

The final step is a key monitoring step. In this step, the results of the changes are to be continuously monitored by the IE Lab team.

REFERENCES

Akst, D. (2013). What can we learn from past anxiety over automation. *Wilson Quarterly*, Summer. Woodrow Wilson Center, Washington, DC.

Autor, D. H., Levy, F., and Murnane, R. J. (2003). The skill content of recent technological change: An empirical exploration. *Quarterly Journal of Economics* 118(4): 1279–1333.

Autor, D. H., Dorn, D., and Hanson, G. H. (2013). The China syndrome: Local labor market effects of import competition in the United States. *American Economic Review* 103(6): 2121–2168.

Autor, D. H. (2015). Why are there still so many jobs? The history and future of workplace automation. *Journal of Economic Perspectives* 29(3): 3–30.

Brynjolfsson, E., and McAfee, A. (2014). *The Second Machine Age: Work, Progress, and Prosperity in a Time of Brilliant Technologies*. New York: Norton.

Frey, C. B., and Osborne, M. A. (2013). The future of employment: How susceptible are jobs to computerization? Oxford Martin School, September.

Hammer, M., and Champy, J. (1993). *Reengineering the Corporation: A Manifesto for Business Revolution*. New York: Harper Business; revised updated ed., HarperCollins, 2004.

Kanalley, C. (2011). Watson's final Jeopardy blunder in day 2 of IBM challenge. *Huffington Post*, February 15.

PART III

REAL WORLD CASE STUDIES

CHAPTER 5

ACTIVE ADVISING WITH INTELLIGENT AGENTS

5.1 INTRODUCTION

In this chapter, we describe an intelligent machine for the wealth management advisory market. The solution we offer will transition this industry to Enterprise 4.0. This industry is already witnessing the first innings of a major transformation due to a number of factors.

We first describe the business and current practices and then the RAGE AI™ solution that will assist in transforming this industry to the E4.0 state. This solution is already in use in global wealth management firms.

5.2 THE INVESTMENT ADVISORY MARKET

The investment advisory services market is undergoing significant change driven by a multitude of factors. On the one hand, increased regulatory compliance requirements are raising the costs of doing business and putting pressure on business model of asset and wealth management firms. On the other, the rapidly evolving technology environment is enabling new disruptive business models ("robo-advisors") that threaten to commoditize the "high-touch" advisory model practiced by asset and wealth managers. Simultaneously,

The Intelligent Enterprise in the Era of Big Data, First Edition. Venkat Srinivasan.

finance theory is evolving to recognize that there is a bridge between a strictly behavioral approach and a strictly mean-variance-based framework in terms of optimizing an individual's wealth.

Optimal design of portfolios has been the subject of much attention and discourse in the academic literature. Recently, behavioral finance and the conventional mean-variance framework have been shown to converge with an appropriate definition of risk [1, 4, 5, 8, 12, 13, 16]. The key implication is the recognition that we need to focus on people's risk behavior at different levels of wealth in order to devise an optimal portfolio design. A corollary implication is the need to optimize the net worth of the individual versus only optimizing assets.

Enabled by direct accessibility to investors as a result of the Internet and rapid advances in technology, robo-advisors have emerged as a disruptive business model at the retail end of the wealth management market. Their attractiveness is a significantly lower fee compared to the traditional high-touch wealth management firms.

High net worth individuals [HNWI] advising is considerably more complex compared to the retail market. Each client has idiosyncratic needs driven by a set of personal factors. Financial advisors for these clients employ highly personalized services to meet their needs. On top of the idiosyncrasies of the HNWI market, wealth advisors (WAs) have to understand a whole new generation, Millennials. Millennials already control a significant amount of wealth and there is a massive wealth transfer from Baby Boomers to Millennials in the United States.

While there is no denying the increased level of complexity as one moves up the wealth spectrum, the current advisory model requires a significant amount of touch and does not leverage technology effectively. This is even more important to be able to address the needs of Millennials who are technology friendly. Robo-advisors are demonstrating that currently there is not even effective leveraging of technology for retail, supposedly simple, clients.

The number of WAs who follow the optimal portfolio design approaches recommended in the recent literature are few and far between. Most WAs are still focused on historical performance and don't spend enough time thinking about systematically managing client needs and preferences proactively. This is largely because of the lack of a holistic technology framework to enable them to reflect their accumulated knowledge and experience for the benefit of their clients.

The rapid advances in technology have created opportunities for altering the depth and breadth of technology leverage in delivering advisory services. As described in Chapters 2 and 3, this industry can and will transition to the intelligent enterprise state. We outline such an intelligent system and framework in this chapter.

5.3 WHAT DO INVESTORS REALLY NEED AND WANT

A lot of advisors and investors simply follow the standard doctrine of creating a mean-variance efficient portfolio. However this does not completely capture investor behavior. As Statman (2004) points out, this only serves the utilitarian benefits for an investor and not their expressive or emotional benefits. Most investors exhibit two emotions when evaluating risky choices: fear and hope. Lopes (1987) developed a framework for choices under uncertainty based on this observation. On the other hand, it is well established (Das, Scheid, and Statman, 2010) that people simplify choices by dividing joint probability distributions into different mental accounts or buckets.

Shefrin and Statman (2000) combine the Lopes *SP/A* (S = security, P = potential and A = aspiration) framework with the mental accounting framework of Kahnemann and Tversky (1979) to develop a behavioral portfolio theory (BPT). Recognizing that most people will have multiple mental accounts (or buckets) of risk and aspiration, BPT provides a framework for designing optimal portfolios for each mental account (level) or bucket. They argue that such optimal design choices may not be strictly compatible with a pure mean-variance efficient portfolio.

The idea that people think differently at different wealth levels is universally valid. Equally valid is that people at all wealth levels have fears and aspirations. Thus, from a wealth management perspective, an appropriate framework would require the WAs to understand what the investors really need and want at various levels of wealth suited to their current state.

Chhabra (2005, 2015) and Chhabra and Zaharoff (2001) are widely credited with creating a goals-based wealth management framework based on integrating the principles of behavioral finance and the mean-variance framework. In the *Aspirational Investor*, Chhabra (2015) provides a lucid argument for such a framework and an idealized example to illustrate its application.

5.4 CHALLENGES WITH HIGH-TOUCH ADVISORY SERVICES

5.4.1 Questions of Value and Interest

In the financial crisis of 2008 we reached a low mark in people's trust for the financial services industry in general. The private banking and wealth management industry clearly depends on its trusted adviser status to function and prosper. The crisis brought to the fore real questions of value and conflict that have always existed. Where is the focus on client outcomes? What value does a high-touch WA bring with the significant fees that comes with it? These have become central questions for the industry.

Prior to the focus on fees and value, there were questions of conflict with proprietary funds. The industry has by and large addressed that by moving to an open architecture model.

There is a real need for WAs to understand their clients better. A recent PwC survey (2013) finds, in this regard, that while survey respondents were clear that many aspects of their client relationships are effective, key demographic trends (including the rise in importance of Generation Y and of women as specific client segments), now need to be embedded within segmentation techniques, next generation transition management needs to improve, and profitability measurement needs to become more sophisticated. Respondents ranked themselves as needing to improve in some key areas of client value added, especially in client reporting, digital offerings, and provision of broad-based advice. The industry needs to become smarter at understanding what clients really value, and in turn how much they will pay for the value added. However, the survey also reports that private clients the survey included were currently dissatisfied by how they were communicated with by their wealth managers.

WAs including Registered Independent Advisors [RIAs] face a real challenge in addressing the value versus fee question. Established full-service firms in turn have a significant challenge to support the costs of a high-touch model. They face significant margin pressures. The emphasis is now on solutions and a shift away from perceived commoditized products. The industry value chain is evolving with greater specialization and focus on the key determinants of success.

5.4.2 The Massive "Wealth Transfer" Phenomenon

Every generation has seen a shift in its overall attitudes and values largely influenced by the major events that generation has witnessed. In the United States alone, there are approximately 80 million Millennials born between 1980 and 2000, and the oldest in the generation are entering their 30s. This generation's views and attitudes have been shaped by rapid globalization and the pervasive influence and advances of technology on all aspects of life.

Millennials already control approximately $2 trillion in liquid assets, and that is expected to surge to $7 trillion by the end of the decade. Over the next several decades, it is estimated that $30 trillion in assets will transfer from Baby Boomers to their heirs in the United States alone. The transfer is expected to peak between 2031 and 2045 when 10% of total wealth in the United States will be changing hands every five years (Accenture, 2012, 2013).

It is widely reported that Millennials are widely distrustful of financial markets, consider social responsibility to be a major factor in investments, believe in their ability to change the world, and a majority of them declare themselves as "self-directed" investors. They believe in "doing well by doing good." WAs cannot expect that the Millennials will simply go along with their parents' view of the world and investment principles. WAs have to be in sync with a very different outlook and approach on the part of their clients to wealth.

5.4.3 The Rise of Robo-Advisors

Since 2009, more than 200 companies have jumped into the business of helping investors with their asset management portfolio management online. These range from venture-capital-backed start-ups to established firms like Fidelity Investments, Vanguard Group, and Charles Schwab. These companies are attempting to disrupt the traditional wealth management market using automation for portfolio selection and low or even no fees for their services.

Investors can pick from pre-constructed portfolios or build their own portfolios with easy to use tools. Some firms, including *SigFig Wealth Management* and *FutureAdvisor*, offer a free analysis of an investor's existing holdings and recommend ways to change portfolios to achieve an optimal mix of securities. For an annual fee of 0.25% and 0.5%, respectively, *SigFig* and *FutureAdvisor* will also manage those assets. Based on a qualitative online understanding of the risk profile of the investor from a questionnaire, they automatically recommend an allocation of assets from a relatively static set of options largely based in index and exchanged traded funds (ETF) based investing.

Should established firms be worried about these nimble, disruptive business models? How should they compete with such businesses? On the flip side, what do these new businesses have to do to effectively service higher wealth segments?

Not surprisingly, the paths for both are similar. Established firms have to deploy much more flexible technology if they aspire to compete with these new businesses. Robo-advisors have to do the same to be able to move up the market to serve higher level wealth segments.

5.4.4 Technology for HNWI's Unique Needs

Conventional wisdom has been that the HNWI market is complex, with each client having their unique and idiosyncratic needs and thus requiring high-touch services. While it certainly cannot be served by "one size fits all" automated services, it is also true that many WAs are not leveraging all the available technology to offer their services in a consistent and systematic fashion to HNWI clients.

Regardless of any competitive need from robo-advisors, established firms have an opportunity to adopt flexible technology that can help them address HNWI needs much more effectively. They can use technology to rapidly configure the needs of HNWIs at a very granular level, even at the level of a family member, without it costing an arm and a leg. They should be able to reflect the diverse aspirations and values of their clients including Millennials.

Adopting next generation technology will enable WAs to demonstrate how they can add value to their HNWI clients. Today, many WAs simply do not have continuous visibility into the state of their client's net worth and what is likely to impact it. They have mostly static views and produce voluminous reports of historical performance of their client's investment portfolios that are of limited or no value.

They should be able to systematically monitor the external environment and map changes in the environment with client needs proactively. Today, it is largely an episodic activity driven by the WAs memory and ability to connect the dots.

5.5 ACTIVE ADVISING – A FRAMEWORK BASED ON MACHINE INTELLIGENCE

This industry is ripe for a major transformation as a result of both technology advances and other factors explained above. We would characterize the current state of this industry somewhere between E2.0 and E3.0. Many of the largest wealth management firms do not even aggregate holdings from multiple custodians, have limited or no flexibility in addressing their client needs, and the advisor community has to work outside the system to implement their value-added ideas.

Moving this industry to E4.0 is quite feasible and inevitable. Wealth advising is a classic knowledge-based process that can be supported with intelligent machines. Different aspects of this process require different kind of intelligent machines. We describe LiveWealth™, an intelligent system based on RAGE

AI™, which can non-intrusively transition any wealth management firm or advisor to E4.0.

The wealth management industry services clients across a wide range of the wealth spectrum with different needs and desires. The industry is generally divided into the following segments: retail, high net worth, and institutional clients. The high net worth segment may be further divided into high net worth and ultra high net worth segments. For the retail segment, as the robo-advisors have demonstrated, wealth management actions can be highly automated and require very little or no human touch. However, Robo-advisors generally have created very specific and limited systems that reflect a very deterministic view of what retail investors want and need. They are in E3.0. They demonstrate little or no flexibility very specific, limited options, and certainly no level of machine intelligence. With LiveWealth™, wealth management firms can define the level of automation that is most appropriate for the client and don't have to arbitrarily limit choices.

For WAs to demonstrate true value, they have to become "active advisors." They have to be constantly aware of what their clients value and care about, continuously monitor changes in the external environment that can impact their client's wealth, and changes in the client's personal environment that can impact the client's preferences and objectives. They have to take an "active" view of advising and not a static, periodic, or an episodic view.

Many successful WAs attempt to be "active" through sheer effort. They try and stay on top of all the developments they think could impact their clients. Given the amount of information and the factors that can impact a client's wealth position, they are not likely to be consistently successful. Besides, it is a very taxing and unsustainable model.

Figure 5.1 displays a high-level architectural view of LiveWealth™ (LW), an intelligent wealth management advisory system from RAGE. LW turns every advisor into an active advisor. The RAGE AI™ platform has been used to create intelligent agents that can function at the behest of the WA. They monitor both structured and unstructured data using the WAs logic, when needed, and inform the WA as desired. The intelligent agent framework comprises an extensive set of intelligent agents that are highly configurable. They can be configured for each client or member of the client's family. They can also be configured to reflect the WAs expertise in analyzing performance data and other developments.

These intelligent agents are automated robots that monitor the external and internal environments continuously. They draw the WAs attention to trends and developments that need action. We will describe a few of these agents here.

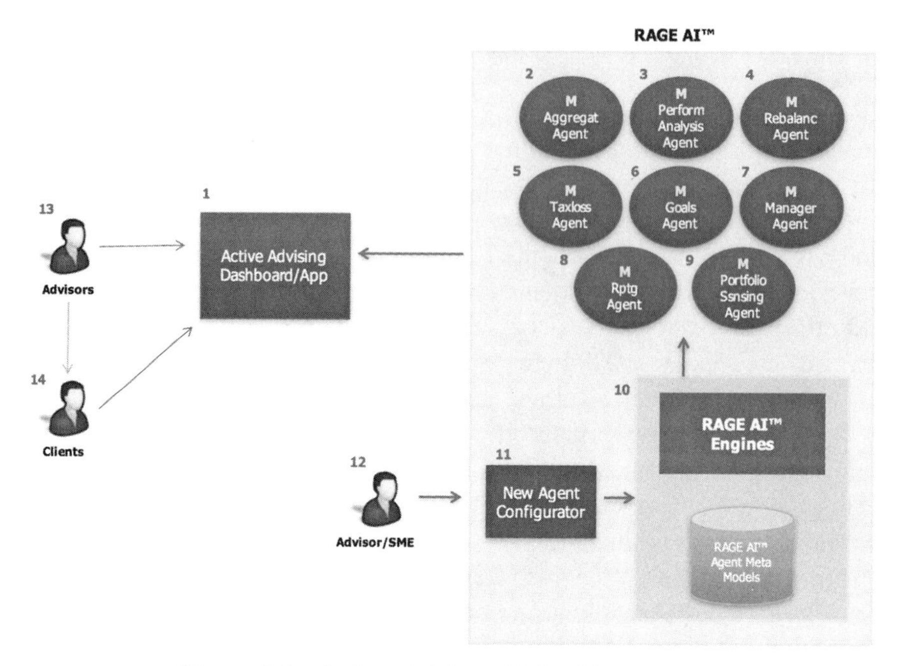

Figure 5.1 Active advising with intelligent agents

5.6 A HOLISTIC VIEW OF THE CLIENT'S NEEDS

Central to LW is a holistic understanding of the client's needs. This can be done through an appropriate set of questions and a categorization of the client's situation and wealth objectives. LW allows the WA to set up as many categories of goals and for as many members of the family as needed. Figure 5.2 shows a sample Household Balance Sheet organized by goals in three buckets – Needs, Wants, and Wishes.

Needs reflect the basic living expenses of the client and should be appropriately invested. Note the matching of assets and liabilities at each goal level. Wants are the next level of wealth aspiration, and finally, Wishes are aspirational goals of the client. LW allows the WA to assess the client's risk tolerance though a series of questions. These questions can be configured for each client to reflect their unique situation.

As noted earlier, LW enables the WA to setup any number of goal levels for the client. The WA can even setup separate levels for different members of the client's household if necessary.

The *goals assessment and monitoring agent* (Figure 5.3, 5.4) continuously reviews and checks the client's goals, actual performance, and the potential gaps. When the goals are set up initially, the agent does a quick assessment

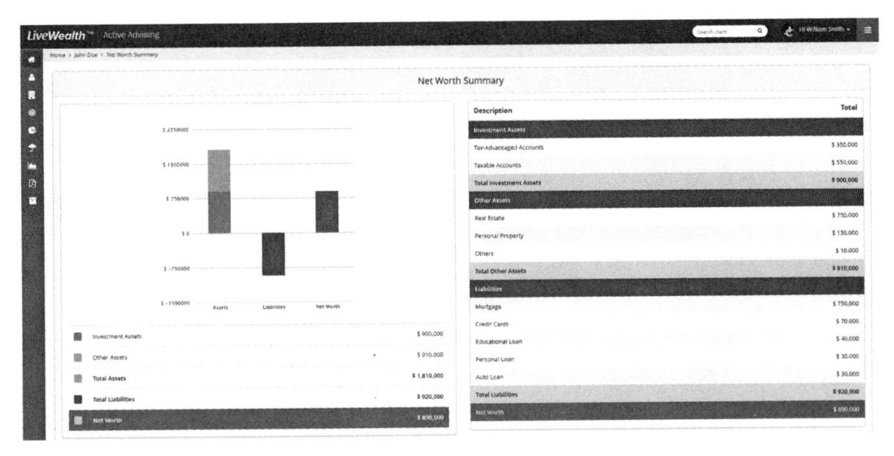

Figure 5.2 Household Balance Sheet

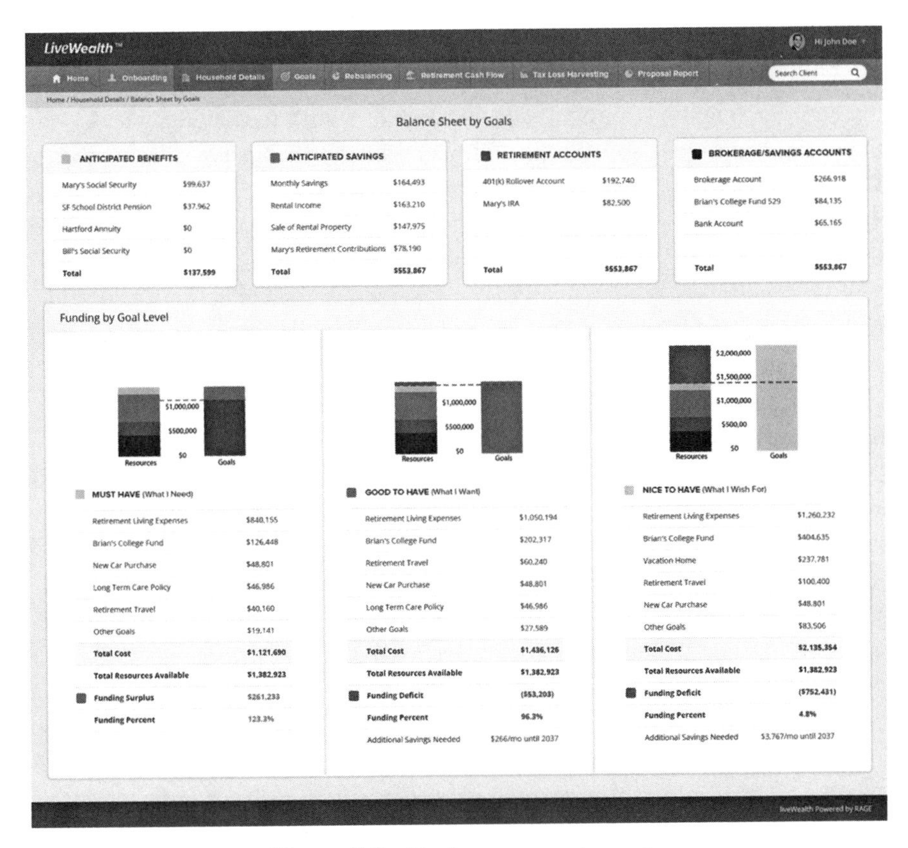

Figure 5.3 Goals assessment agent

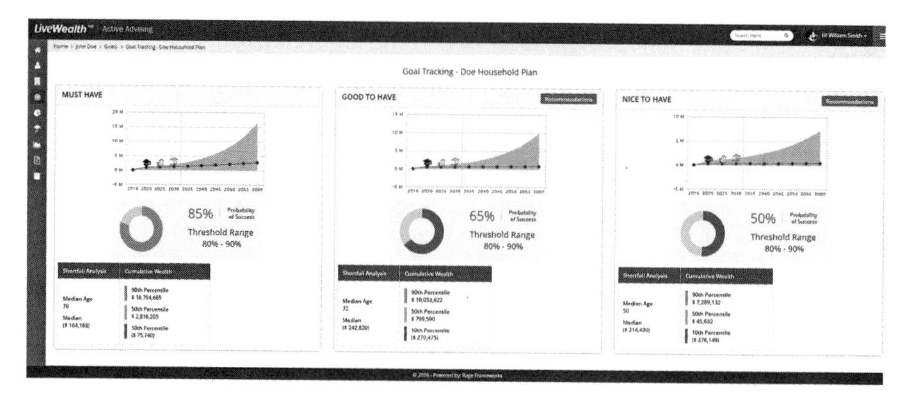

Figure 5.4 Client risk tolerance

of their feasibility. After the setup, it monitors them continuously. If it detects that the actual performance is deviating or likely to deviate from the target for any or the goals, it is also capable of determining possible actions to fill the gap. Such actions may include suggesting a shift to a higher yield, additional contributions, and the like. Built in simulators in LW also allow the WA to simulate unexpected shocks and events in the client's environment. This enables the WA to understand the impact of such shocks and to plan for them appropriately. The WA is alerted via the Active Advising dashboard. The WA can either accept and approve the recommended actions or override them and define revised actions to address the gap between the client's current position and their goals.

The *aggregation agent* automatically aggregates all the assets and liabilities of the client from one or more sources. It is, of course, more valuable in the case of clients who have multiple accounts with different firms – custodians, brokerages, and the like. Historically, most advisors even in many of the largest firms have not had the technology to aggregate all the data from these multiple accounts at a transaction level easily. Suppose that you have an advisory relationship with JP Morgan Chase, a brokerage account with Fidelity, and your 401k with Charles Schwab, your advisor at JP Morgan Chase needs to aggregate all of your assets (and liabilities) before coming up with a suitable financial plan. Today, there is no easy way for this advisor to automatically aggregate all this information. There are no standards or standardized process for sharing information in the industry. Many advisors do it manually on spreadsheets based on paper statements or PDF documents they obtain from their clients. Some of these statements can run into hundreds of pages. You can imagine how messy and error prone this process must be!

The aggregation agent solves this challenge. Regardless of the format and type of document, the aggregation agent extracts the relevant information with 100% accuracy and 100% audit trail. RAGE AI™ components intelligently recognize a custodian, brokerage, financial, or bank statement. They automatically extract the data and normalize the data across the statements. The platform's natural language component categorizes and interprets the extracted data – for example, the labels of transactions so that they are interpreted to identify the security, its characteristics, and other attributes like coupon (for a fixed income security), maturity date, and asset class.

Aggregation is an interesting example of knowledge-based automation and involves both extraction and natural language processing. It is also an example where *a black box will not work* because of the auditability requirement. Because of the sensitive nature of the problem and also regulatory requirements to support billing, for example, all the aggregated data and their categorization must have complete support.

The *rebalancing agent* monitors portfolio values continuously against a target allocation and, based on thresholds set by the WA, identifies if rebalancing action may be timely. The agent also determines the set of actions necessary to rebalance.

In addition, if the WA provides the appropriate parameters, the rebalancing agent can also identify a set of recommendations that are the most appropriate for the client. This can take into account current performance of funds, or changes in any client circumstances including any shifts in goals and liabilities, and then applies WA defined rules to identify recommendations for rebalancing actions.

Figure 5.5 shows a partial view of the rebalancing agent. It is a live process not just a flow chart. The WA can configure the rebalancing agent for each client and even within each client at more granular levels. Figure 5.6 shows a sample live rebalancing process implemented for a specific client.

The *goals monitoring agent* specifically monitors the relevance of client goals and how well the client's actual positions, both assets and liabilities, are tracking to the plan set by the WA. It can factor such scenarios as a possible change in interest rates in the future or change in real estate values, in addition to actual values of tradable asset portfolios. Figure 5.7 shows the results of the goals monitoring agent for a sample client at a point in time. The pie charts in Figure 5.7 indicate the current feasibility of the goal.

If configured to do so, the agent can automatically request for an online update on the personal circumstances of the client on which the WA has relied on to create a wealth-maximization strategy. Most WAs in the HNWI market will prefer to do this in person with their client. The goals monitoring agent

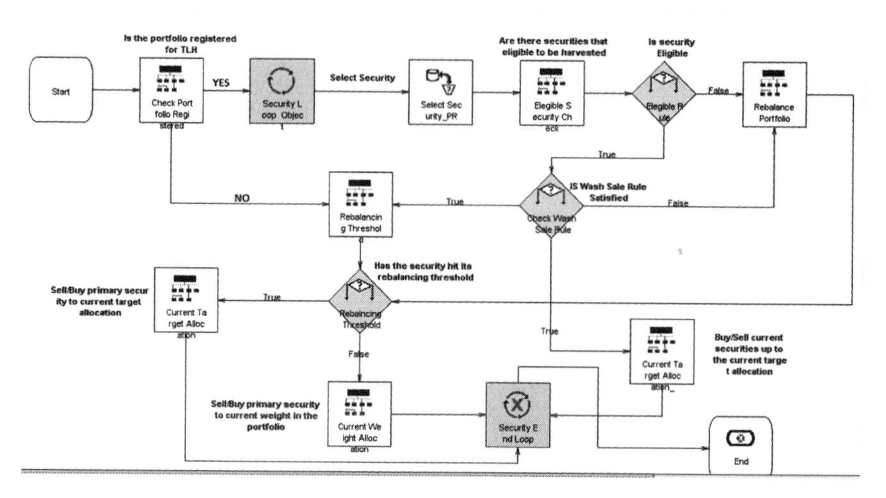

Figure 5.5 Rebalancing agent

Figure 5.6 Rebalancing agent results view

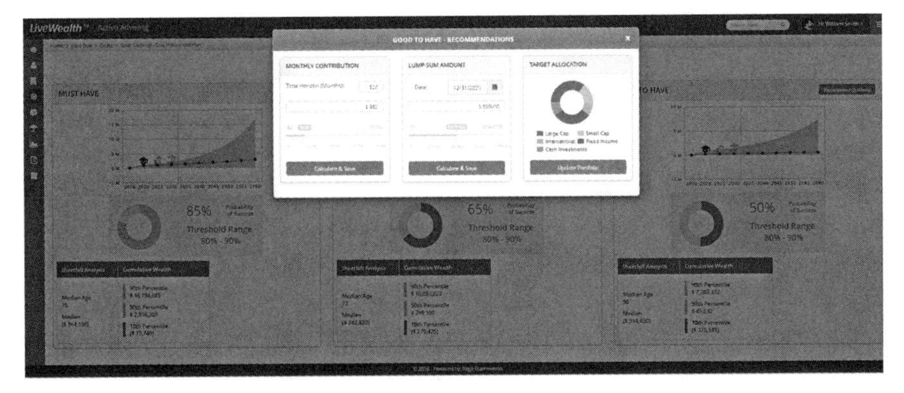

Figure 5.7 Goals monitoring agent

can, however, let them prepare for their meeting with the client with intelligent What-If scenario building.

The *tax planning agent* can implement tax planning strategies including the WAs tax-loss harvesting logic, if any. The WA can potentially set up client-specific tax-loss harvesting rules/logic. The tax planning agent can go far beyond just tax-loss harvesting and implement complex tax efficient wealth management techniques. Figure 5.8 shows a sample tax loss harvesting agent and the results screen.

The *liability optimization agent* can monitor the client's liabilities, the external environment, and trigger opportunities for optimizing. This is particularly appropriate for a multi-line financial institution.

The *portfolio sensing agent* monitors the external environment for developments that can impact a client's portfolio. Today, the WAs have no systematic way of monitoring the multitude of external events that can impact their client's wealth or events that their clients care about. The amount of digital information at our disposal is so huge that it is humanly impossible for us to be on top of all of it. Figure 5.9 displays a heat map to the WA

The sensing agent addresses this information overload problem by using machine learning and natural language processing methods to process the millions of external events against a set of relevant factors for each client to identify the events that should be of interest. The WA can configure such a factor list, which we refer to as a semantic ontology.

The sensing agent can quickly detect the impact of developing events such as a news item that carries an announcement by the governor of Puerto Rico that the state cannot afford its debt burden. It can identify all the clients that are

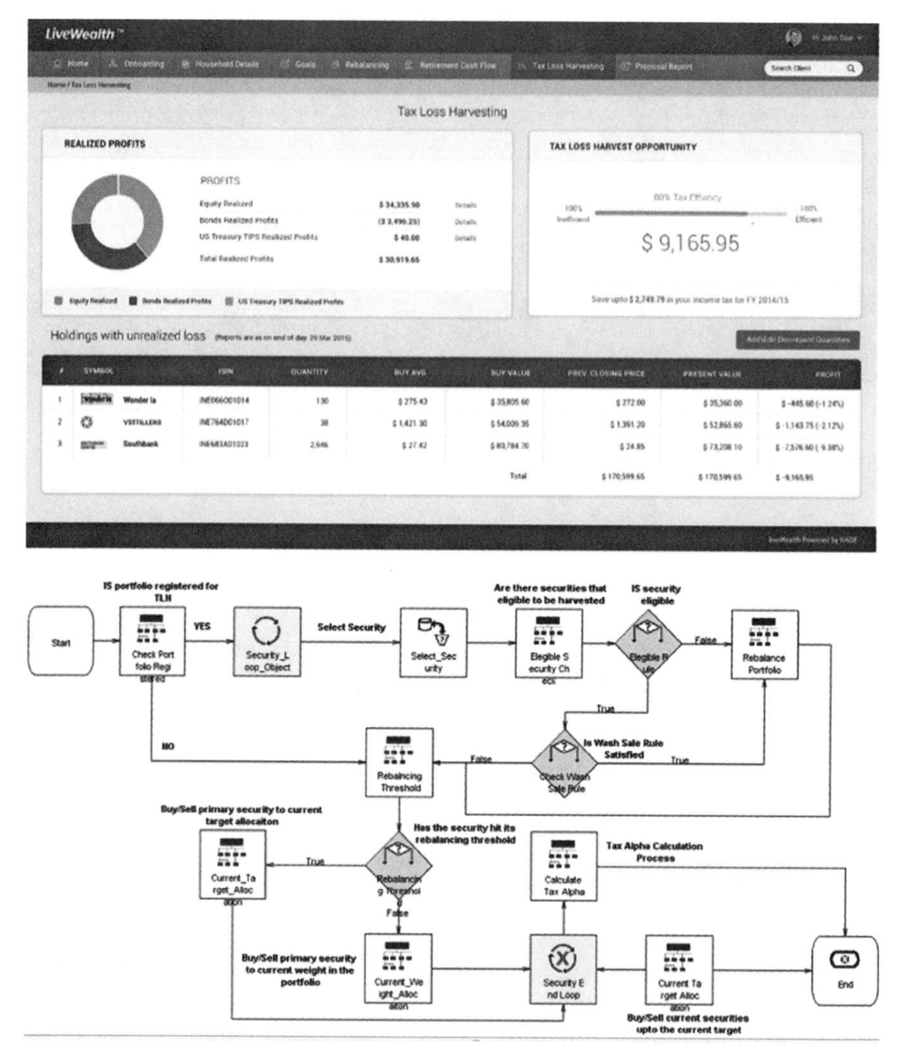

Figure 5.8 Tax-loss harvesting agent

holding Puerto Rico municipal bonds and quantify the magnitude of exposure of various clients to Puerto Rico bonds and also which clients goals are likely to be impacted by what magnitude.

The sensing agent updates a heat map for each client (or more granular entities). This enables the WAs to be much more proactive in managing their client's wealth.

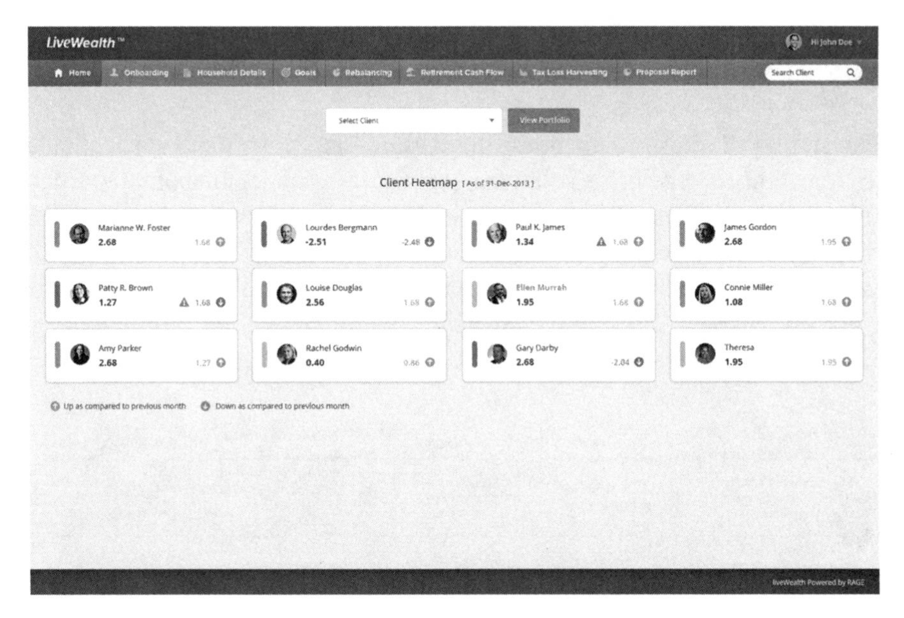

Figure 5.9 Sensing agent heat map

5.7 SUMMARY

The $30+ trillion wealth management market is undergoing significant changes as a result of major macroeconomic shifts in the world, demographic changes with respect to transfer of wealth to a new generation, and a set of regulatory changes impacting the industry. At the same time, technology has evolved and created opportunities for radical transformations of business models in the industry.

High-touch advisory models can be significantly transformed to reduce the level of touch and deliver high value to clients. Technology can intelligently automate routine tasks in a very flexible manner and can help the wealth manager be more proactive even in knowledge-based tasks. Overall, these will enable the WA to become an "active advisor."

We have described an intelligent system that can transition wealth management firms and advisors in E4.0. The system has the intelligence to intelligently implement the policies and strategies that the WAs design. It can also automatically access domain-specific facts, ranging from custodian statements to sensing what developments in the external environment are likely to affect clients.

APPENDIX: THE RAGE BUSINESS PROCESS AUTOMATION AND COGNITIVE INTELLIGENCE PLATFORM

Powered by 20 abstract engines, the RAGE Enterprise Platform facilitates the creation of enterprise solutions rapidly in a matter of hours and days (Figure 5A.1). The platform has a unique model-driven architecture that reduces software development to a modeling exercise.

Using the RAGE platform, we have created the Intelligent Agent framework to enable Active Advising Figure 5A.2). Agents are intelligent

BUSINESS RULES ENGINE	MODEL ENGINE	NLP ENGINE	PROCESS ASSEMBLY ENGINE	WEB SERVICES ENGINE
Define business logic as a function of application information	Define computational expression/algorithm	Processing to detect patterns in unstructured text	Process aids such as sequencing, looping, branching, synchronizing	Easy assembly of Web Services for other applications
DECISION TREE ENGINE	**MODEL NETWORK ENGINE**	**COMPUTATIONAL LINGUISTICS ENGINE**	**DATA ACCESS ENGINE**	**DESKTOP INTEGRATION ENGINE**
Chain a collection of business rules and desired outcomes	Define a network of models which may be connected through business rules	Linguistics to interpret English language text in context	Access data in the application database	Integrate Microsoft products, e.g., Word, Excel, Powerpoint, as part of any process
CONNECTOR FACTORY ENGINE	**QUESTIONNAIRE ENGINE**	**REAL TIME CONTENT INTEGRATION ENGINE**	**ASSIGNMENT ENGINE**	**MESSAGE ENGINE**
Create connectors between systems	Dynamic Best Next Action	Integrate RSS feeds, acquire content through crawling	Create/modify process information	Create mail messages as part of the process
USER INTERFACE ENGINE	**EXTERNAL OBJECTS ENGINE**	**EXTRACTION ENGINE**	**REPOSITORY**	**INTLLIGENT DOCUMENT BUILDER ENGINE**
Rapid assembly of fully functional user interfaces integrated into process flow	Integrate external programs as tasks in any process	Extract data from unstructured documents	Hold Application Level Meta Data	Create intelligent narrative and documents

Figure 5A.1 Business process automation (BPA) and cognitive intelligence (CI) platform

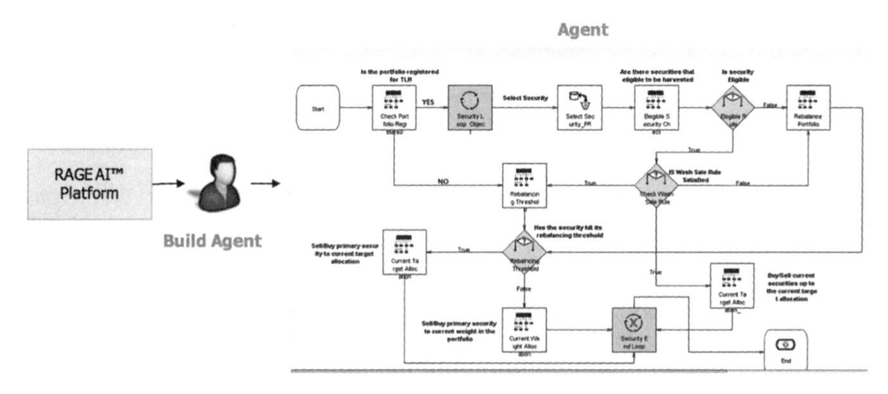

Figure 5A.2 Intelligent agent framework

automation processes that can be created on-demand at any level of granularity that the WA desires – client, portfolio, and so on. They can handle structured data like performance returns, or unstructured data like news.

REFERENCES

Accenture (2012). The "greater" wealth transfer: Capitalizing on the intergenerational shift in wealth, June. https://www.accenture.com/usen/~/media/Accenture/Conversion-Assets/DotCom/Documents/Global/PDF/Industries_5/Accenture-CM-AWAMS-Wealth-Transfer-Final-June2012-Web-Version.pdf.

Booz & Co. (2008). *U.S. Wealth Management Survey*. London, U.K.

Brunel, J. (2012). Goals-based wealth management. CFA Institute, March: 57–65.

Brunel, J. (2003). Revisiting the asset allocation challenge through a behavioral finance lens. *Journal of Wealth Management* 6(2): 10–20.

Chhabra, A. (2015). *The Aspirational Investor*. New York: Harper-Collins.

Chhabra, Ashvin B. (2005). Beyond Markovitz: A comprehensive wealth allocation framework for individual investors. *Journal of Wealth Management* 7(4): 8–34.

Chhabra, A., and Zaharoff, L. (2001). Setting an asset allocation strategy by balancing personal and market risks. *Journal of Wealth Management* (Winter): 30–33.

Das, S., Markovitz, H., Scheid J., and Statman, M. (2010). Portfolio optimization with mental accounts. *Journal of Financial and Quantitative Analysis* 45(2): 1–24.

Kahnemann, D., and Tversky, A. (1979). Prospect theory: An analysis of decision making under risk. *Econometrica* 47: 263–291.

Lopes, L. (1987). Between hope and fear: The psychology of risk. *Advances in Experimental Social Psychology* 20: 255–295.

Markowitz, H. (2002a). Portfolio selection. *Journal of Finance* 7: 77–91.

Markowitz, H. (2002b). Investment for the long run: New evidence for an old rule. *Journal of Finance* 31: 1273–1286.

Marston, C. (2014). Great wealth transfer will be $30 trillion – yes, that's trillion with a T. CNBC, http://www.cnbc.com/2014/07/22/great-wealth-transfer-will-be-30-trillionyes-that's-trillion-with-a-t.html.

Merrill Lynch (2013). Millennials and money. http://www.pbig.ml.com/pwa/pages/Millennials-and-Money.aspx

Moyer, L. (2015). Putting Robo-advisors to the test. *Wall Street Journal*, April 27.

O'Brien, S. (2015). Robo firms gaining traction with traditional advisors. Special to CNBC.com, April 17.

Osterland, A. (2014). Robo-advisors are no immediate threat, wealth managers say. NBC.com, June 23.

Price Waterhouse Coopers (2013). Private Banking and Wealth Management Survey, http://www.pwc.com/gx/en/banking-capital-markets/private-banking-wealth-management-survey/download-the-report.jhtml

RAGE Frameworks (2013). A large scale system for contextually relevant analysis of unstructured content. Rage White Paper, Dedham, MA.

Sharf, S. (2014). The Recession Generation: How Millennials are changing money management forever. *Forbes*, July 30.

Shefrin, H., and Statman, M. (2000). Behavioral portfolio theory. *Journal of Financial and Quantitative Analysis* 35(2): 127–151.

Shiller, R. J. (2001). *Irrational Exuberance*. Princeton: Princeton University Press.

Statman, M., (2004). What do investors want? *Journal of Portfolio Management*, 30th Anniversary Issue, 153–161.

Tversky, A., and Kahnemann, D. (1992). Advances in prospect theory: Cumulative representation of uncertainty. *Journal of Risk and Uncertainty* 5: 297–323.

CHAPTER 6

FINDING ALPHA IN FINANCIAL MARKETS

6.1 INTRODUCTION

Active managers (mutual funds, hedge funds) are constantly looking to find inefficiencies in the markets. How do they find such inefficiencies? They use a variety of strategies and factors to construct their portfolios. All such strategies and factors are based on the understanding and insight of their research staff using the information they are able to process.

The effectiveness of an active management strategy clearly depends on the skills of the manager and the research staff. The conventional wisdom based on the The Standard and Poor's Index Versus Active (SPIVA) quarterly scorecards was that only a small number of actively managed mutual funds have gains better than the SPIVA benchmark. However, recent research (Brooks, 2001) shows that actively managed funds outperform benchmarks if we were to look more closely at the active portion of these portfolios. This suggests a clear informational value.

A recent historical study of active versus passive investing (Hartford Funds, 2015) points out that active management has typically outperformed passive management during market corrections. The study argues that active and passive management have moved in cycles. In certain cycles with high levels of

The Intelligent Enterprise in the Era of Big Data, First Edition. Venkat Srinivasan.

dispersion and volatility, active managers have tended to outperform passive managers. In periods with low levels of dispersion, vice versa has been true.

We believe, however, that today's active manager is handicapped by an inability to process the explosive amount of information that is continuously arriving in the market. These studies are not able to account for whether there are true informational inefficiencies in the market that the active managers are able to exploit.

In today's digital world, we believe the effectiveness of any active investment strategy critically depends on the ability to process the enormous amount of information that is constantly arriving digitally. We believe that the rate of arrival of new information far outstrips the capacity of investment managers and their staff to process such information. Ignoring any differential capability among active managers to interpret information, it is a bit of a hit or miss in terms of the managers reviewing all relevant information in a timely manner. Sometimes they get to review most of the relevant information and other times not.

In our view, even the most systematic active managers and hedge funds are struggling to process all the information that comes their way. We would characterize the active investment industry between E2.0 and E3.0. This chapter describes the way to transition to an intelligent state.

6.2 INFORMATION ASYMMETRY AND FINANCIAL MARKETS

In Chapter 3, we described various types of information asymmetry that results in commercial transactions. In the context of financial markets, Fama (1970) defines an efficient market as a market where assets values always fully reflect all the available information, the efficient market hypothesis (EMH). The assumption underlying EMH is that when new information concerning the fundamental value of an asset arrives in the market, its price should react instantaneously, which is to say, new information should diffuse immediately and perfectly. Of course, superior or differential access to information can yield superior returns (Figure 6.1).

There is also considerable literature on the impact of investor sentiment on asset prices. DeBondt and Thaler (1985) form portfolios of the best and worst performing companies over the past three years and compute the corresponding returns over the following five years and find that while long-term trends tend to reverse, short-term trends tend to continue. Many studies have reported similar findings arguing that there is under- and overreaction in financial markets (Zarowin, 1989; Cutler et al., 1991; Bernard, 1993; Chan et al., 1996; LaPorta, 1996; LaPorta et al., 1997; Daniel et al., 1998; Hong and Stein,

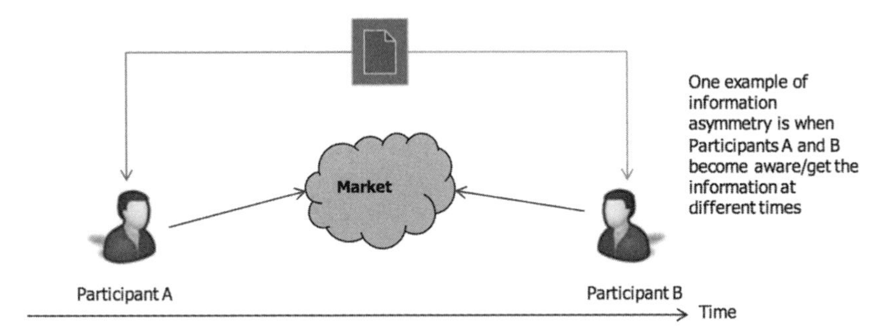

Figure 6.1 Information asymmetry

1999; Barberis and Shleifer, 2003; Kadiyala and Rau, 2004; Zhang, 2006; Moskowitz et al., 2012; Stambaugh et al., 2012). These studies have collectively rationalized underreaction with slow diffusion of news, conservatism, the disposition to sell winners too early and hold on to losers too long (Shefrin and Statman, 1985; Frazzini, 2006; Barberis and Xiong, 2009). Similarly, overreaction is argued to be because of positive feedback trading, investors' overconfidence, representative heuristics (Tversky and Kahnemann, 1974; Barberis et al., 1998), and herding behavior (Bikchandani et al., 1992).

Miller (1977) argues that it is implausible to assume that, although the future is very uncertain and forecasts are very likely to miss, that somehow everyone makes identical estimates of the return and risk from every security. In practice, the very concept of uncertainly implies that reasonable people may differ in their forecasts. There is now considerable doubt in the literature on whether the EMH can always endure in the real world, as investor sentiment in the presence of limited arbitrage may be persistent over long time horizons (Shleifer and Summers, 1990; Chen et al., 2002; Hong et al., 2006).

Stambaugh (2012) finds that observing and forecasting sentiment shifts does pay off in financial markets as arbitrageurs track price trends, trading volumes, volatility indexes, investor surveys, and other gauges reflecting sentiment shifts. Some therefore argue that the impact of sentiment on asset prices is an irrefutable fact of life and the only issue is how to measure it (Baker and Wurgler, 2007). In a comprehensive survey of the literature on the topic, Bank and Brustbauer (2014) observe that current research does not shed any light on factors influencing investors' individual evaluation of asset characteristics. They suggest that future research should focus on possible symptoms of investor sentiment that make investors prone to sentiment.

We believe there is an important dimension related to information diffusion and investor sentiment that has not attracted much attention in the literature. Our focus in this chapter is on how to systematically interpret information as soon as it arrives and determine its possible impact on asset prices. In this digital world, where the rate and quantum of information arrival is explosively high, we believe that it is impossible for all information to diffuse throughout the market instantaneously. We examine the possible hypothesis that if we are able to develop superior access to all information and interpret such information without any personality bias (or a constant and nonchanging sentiment bias), we will be able to measure the diffusion of information and its impact on asset prices.

There have been several studies on the rate of information diffusion and the effects of information diffusion on security prices. The majority of these studies have examined returns on or around different types of events, such as earnings announcements.

Several theories have attempted to explain the empirical observation that stock prices tend to drift over time in the same direction as the initial new information. Generally, researchers have examined such drift in the context of earnings announcements and it is referred to as post-earnings announcement drift (PEAD). Three explanations have been suggested for such drift: (1) it may be a risk premium, (2) it may be a result of high transaction costs, and/or (3) it may be a function of the information agents receive. The third explanation is the most relevant to our discussion in this chapter.

There are two competing theories that attempt to explain the differential information argument. On one hand, the "behavioral" theory (Daniel et al., 1998) suggests that the drift is due to the difference between public and private information. On the other, the "rational structural uncertainty" theory (Brav and Heaton, 2002) argues that it is the distribution of information in the economy that matters.

In order to explain the PEAD, Vega (2006) examines which of these two theories best describes empirical data by using variables that distinguish between public and private information, and the arrival rate of informed/uninformed traders. Vega concludes that the drift is not influenced by public or private information. Rather, what matters is whether the information is concentrated. Vega measures public information surprises (SUR) as a combination of the number of times a company is mentioned in the media immediately prior to the event, and daily abnormal stock returns:

$$\sum_{k=2}^{41} SUR_{it} = \{NEWS_{it-k} I(AR_{it-k} > AR_{i20T}) + NEWS_{it-k} I(AR_{it-k} < AR_{i20B})\} \times I(DV_{it-k} < DV_{i50})$$

where $NEWS_{it\text{-}k}$ is a dummy variable equal to one if firm i is mentioned in the headline or lead paragraph of an article on day t and $I(AR_{it\text{-}k} > AR_{i20T})$ and $I(AR_{it\text{-}k} < AR_{j20B})$ are indicator functions equal to one if the abnormal stock return for firm i on day t or day $t + 1$ is above the top 20% and below the bottom 20%, respectively, of daily abnormal stock returns for that firm. $I(DV_{it-k} < DV_{i50})$ is an indicator function equal to one if the detrended turnover for firm i on day $t - k$ or $t - k + 1$ is below the median detrended turnover of firm i.

The study also finds support for the argument that public information sometimes may not be easily interpretable and informed traders who are particularly skillful in analyzing such news may trade on their interpretation (Kim and Verrecchia, 1994, 1997). Chan (2003) finds that firms covered by the media experience larger drift and firms that experience stock price jumps without any observable public news experience returns reversal.

6.3 MACHINE INTELLIGENCE AND ALPHA

We believe that machines can be used to process the vast amounts of information accessible today in a systematic fashion. Active managers can even tailor such machines to reflect their preferences and beliefs if desired. Machines can become active managers' assistants or be operated without any human intervention. As we have argued throughout, such machine learning and intelligence need not be a black box at all.

Let's start with a conceptual model of the impact of information on a firm's intrinsic value. We posit the following hypothesis: a firm is impacted not just by news related to it but any event or information that can impact its operations or business model. This could be, for example, the introduction of a new product by a competitor, or a macroeconomic development that drives demand for the firm's products. In fact, the number of ways a firm can be impacted is large and complex, and requires extensive monitoring/interpretation of the information universe. We can think of the proximity of such information to the firm to be in several hierarchical levels. A first-order information item is one that contains the firm's name, so it can be identified directly with the firm. A second-order item is one that has one degree of separation from the firm. An example of this is an information item about the industry the firm operates in. A second-order item has two degrees of separation, and so on. Anything beyond two degrees of separation is probably a bit too distant for immediate consideration. If it does become relevant, we believe it will reappear within two degrees of separation.

Let $\boldsymbol{I}$ represent the information set that can impact the firm's operations $\boldsymbol{O}$, which in turn is interpreted by investors and reflected in price $\boldsymbol{P}$ of the firm's stock:

$$\boldsymbol{I} \rightarrow \boldsymbol{O} \rightarrow \boldsymbol{P}$$

O is the set of factors that describes the firm's operating model based on the firm's business and industry (see the Appendix at the end of this chapter for an example). The level of information asymmetry and the rate of information diffusion on the stock price can be measured by systematically identifying ***I*** and assessing its impact on ***P*** by calibrating its impact on ***O***. In fact, we believe this is what informed traders attempt to do. Informed traders use a complex information model that must resemble ***O***.

There are practical challenges in the effectiveness with which this happens even with informed traders. With the lack of a systematic way of aggregating, sorting, and assessing the full information set ***I***, and without fully enumerating the operating model ***O***, the success rate of informed traders is generally low. We acknowledge that the completeness of the model above is a function of how complete public information about the firm is, how extensive our access to that public information is, and the completeness of ***O***. We ignore private information based on Chan (2003), and we also ignore emotions.

We have developed a system of the model above based on RAGE AI™ and have been systematically assessing the impact of relevant ***I*** on the corresponding firm's ***O***. The system relies on linguistic technologies to automatically interpret the information, determine its relevance to one or more firms' ***O***, and rate the impact of the information based on the impact phrases contained in the information. The rating is performed using a heuristic rating scale. We have examined the relationship between this impact score, which is a time series, and the price series, to determine if the impact leads price changes. Such a measurement can be defined as a measure of information asymmetry and proxies the rate of information diffusion for that firm. Next, we describe the intelligent system and in subsequent sections the results of applying the system on a large information set.

Our system aggregates, classifies, and interprets large amounts of information from a very large number of global, regional, and local sources to generate potential alpha signals for individual firms. The system uses the RAGE AI™ platform described in Chapter 3. Specifically, the system contains the following key components:

- Firm-specific operating models (***O***), which comprise a network of factors that are likely to impact the firm's intrinsic value.
- RAGE AI™ categorizer, to classify the information into one or more topics and determine its relevance to a firm by examining firm's corresponding operating model. (See Chapter 3 for more details on this categorizer works.)

- RAGE AI™ impact analyzer, to assess the potential impact of the information on the firm by analyzing the collective impact of relevant impact phrases in the document

We provide a high-level architectural overview of the system in Figure 6.2.

To have the ability to interpret with fine-grained contextual relevance and be domain independent, we have developed a framework that we refer to as firm-specific operating models (***O***). The firm-specific operating model has three high-level sections – macroeconomic, sector specific, and firm specific. Each firm in the industry will have firm-specific factors in addition to macroeconomic- and sector-specific factors.

The development of such operating models is automated to a significant extent using the RAGE AI™ knowledge discoverer (KD) described in Chapter 3. Experts seeded the framework with a few basic concepts, and the KD then discovered additional concepts and relationships in an automated fashion by ingesting vast amounts of content, using corporate filings and crawling the public Internet. The result is a causal semantic network of concepts at the company level.

The firm level cognitive semantic network can be reviewed and edited by an analyst or fund manager to reflect their preferences. For the analysis presented later in this chapter, a team of analysts at RAGE provided their preferences. A snapshot of the operating model for the oil and gas industry is listed in the Appendix.

The content to be analyzed is aggregated from a very large number and variety of sources, including Internet URLs, micro-blogging sources like Twitter and Tumblr, and content from news services. The system provides the flexibility of adding new sources (URLs, Twitter users, internal documents, etc.) as data.

The second stage breaks the content into sentences, clauses, and phrases. In this stage, the system also resolves the co-referential and contextual structure of the document. This stage results in a set of clusters that describes the conceptual structure of the document.

A cluster is considered as a set of sentences falling in the same clique. The strength of a cluster C with sentences $s_0, s_1, \ldots, s_L$ is defined by

$$C(s_0, s_1, \ldots, s_L) = \sum_{j=0}^{L} \sum_{i=0}^{L} S(s_i,\ s_j)$$

One can easily identify the central topic of the article by identifying the cluster that has the maximum strength. A document can have multiple clusters

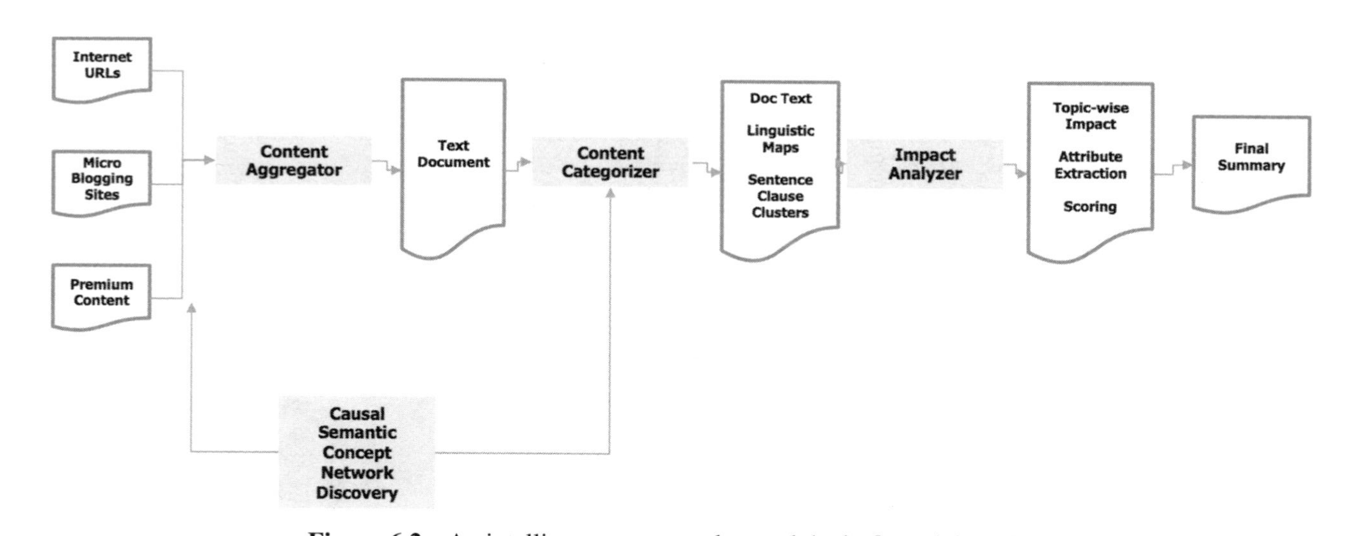

Figure 6.2 An intelligent system to detect alpha in financial markets

and therefore multiple topics. These cluster topics are then overlaid with the causal ontology to categorize the document into a main topic and secondary topics.

In the final stage, the system performs an impact analysis of the document. Impact analysis is done at a phrase level and aggregated at the cluster level. The analysis starts with the main sentence of the cluster and extends through all sentences in the cluster. Impact phrases in the main sentence are identified by a deep linguistic parsing of the sentence and identifying the words and phrases that describe the impact of the sentence.

Impact phrases are much more elaborate in a "meaning" sense than sentiment lexicons used for sentiment analysis. They require a complete linguistic analysis of the discourse and identification of the language structure. Impact phrases are any set of words that communicate impact; they can be verbs, nouns, adjectives, adverbs, or some combination of these in a clause. Impact phrases are an aggregation of primary impact words and words that accentuate the impact. Each clause in a sentence typically communicates impact. Subordinate clauses can modify the impact implied by the main clause, such as in the use "but," which often does just that. Adjectives and adverbs are generally used to accentuate impact; we refer to them as "impact intensity."

In addition to primary impact phrases, the system extracts descriptive attributes like time, quantity, and location. For example, in "Housing starts rose significantly by 3% in January," the verb "rose" is the primary impact word, "significantly" describes intensity, "3%" is also reflective of intensity and "January" is an indicator of time. Such phrases can contain "Negation," as in "Housing starts did not rise in January." Directional reversal can sometimes occur in a more subtle manner; it can also manifest by the use of anti words – "anti-immigration," "unemployment," and so on.

To achieve consistency and comparison across firms and topics, without loss of generality, we normalize the impact assessment to a discrete set of outcomes and associated scores; currently, we use five discrete categories – Significantly positive, Positive, Neutral, Negative, and Significantly negative.

The same impact phrase can have completely opposite interpretation for two topics. For example, "increase in oil prices" will have a +ve impact on oil companies and –ve impact on transportation companies. We enable the configuration of impact direction at a topic/normalized impact level.

This kind of a scoring scheme assigns a N-dimensional score (for N topics in the system/ontology) to a document. This, when viewed as a time series, provides an assessment of the topic/event or entity as a function of time.

6.4 HOW WELL DOES IT WORK?

6.4.1 Data

We have measured the performance of the system on a large corpus of over 100 million content items over the last three to four years against manually curated results. Through the system's data aggregation component, we gathered news items on a real time basis from over 200,000 sources worldwide. Sources include news sources, websites, and blogs. We also have separate results on content from the social media site, Twitter, which are reported separately.

6.4.2 Measuring Lead–Lag Relationship

To test the lead–lag relationship between stock price and RTI the following vector autoregression (VAR) model is estimated by using the ordinary least-squares (OLS) estimation in each equation. Here both the variables *Stock* and *RTI* are endogenous.

$$STOCK_t = \alpha_1 \sum_{i=1}^{k} \beta_i RTI_{t-i} + \sum_{j=1}^{k} \lambda_j Stock_{t-j} + \varepsilon_{1t} \ldots \tag{6.1}$$

$$RTI_t = \alpha_2 + \sum_{i=1}^{k} \gamma_i RTI_{t-i} + \sum_{j=1}^{k} \pi_j Stock_{t-j} + \varepsilon_{2t} \ldots \tag{6.2}$$

where α_1 and α_2 are intercept terms, the coefficients of β_i and γ_i measure the effects of lagged *RTI* and lagged *Stock* on current *Stock*. Similarly, the coefficients of λ_j and π_j measure the effects of lagged *RTI* and lagged *Stock* on current *RTI* and ε are the impulses. If the estimated coefficients on lagged *RTI* in equation (1) are statistically different from zero as a group (i.e., $\sum \beta_i \neq 0$), and the set of estimated coefficients on lagged *Stock* in equation (2) is not statistically different from zero (i.e., $\gamma_i = 0$), then *RTI* leads *Stock*.

Before we estimate the VAR model, we need to examine the stationary properties of the series and to decide the maximum lag length, k. The stationary of the data series is evaluated by the Augmented Dickey–Fuller (ADF) test. The test results are quite sensitive to the lag length. The lag k is decided based on the selection criterion that results in the lowest value – Akaike's information criteria (AIC), Schwarz's criteria (SC), final prediction errors (FPE), and the Hannan–Quinn information criterion (HQ).

Stationarity An economic and financial series that follows a random walk process is called "nonstationary" over time. A nonstationary variable may reach stationarity by differentiating the series t times. The variable is then referred to as an $I(t)$ or integrated of t processes or t unit roots contained in the series. To test unit roots, Dickey and Fuller developed the following regression equation:

$$\Delta Stock_t = \beta_1 + \beta_2 t + \delta \cdot Stock_{t-1} + \sum_{i=1}^{m} \alpha_i \, \Delta Stock_{t-i} + \varepsilon_t$$

where β_1 is a drift, β_2 represents the coefficient on a time trend t, and ε_t is a white noise error term. Here the null hypothesis is that $\delta = 0$; that is, there is a unit root, and the time series is nonstationary. The alternative hypothesis is that δ is less than zero; that is, the time series is stationary. The estimated t value of the coefficient of $Stock_{t-1}$ follows the τ (tau) statistic. If the computed absolute value of the tau statistic ($|\tau|$) exceeds the ADF critical tau values, we reject the hypothesis that $\delta = 0$, in which case the time series can be considered stationary. On the contrary, if the computed $|\tau|$ does not exceed the critical tau value, we do not reject the null hypothesis, which implies that the time series is nonstationary. We can similarly examine the stationary properties for *RTI*.

Co-Integration and Vector Error Correction Model (VECM) If both Stock and RTI are nonstationary in levels and stationary in differences, then there is a chance of a co-integration relationship between them, which reveals the long-run equilibrium relationship between the two series. The Engle–Granger (EG) co-integration test can be applied to investigate the long-run relationship between *Stock* and *RTI*. The co-integrating regression estimated by the ordinary least-square (OLS) technique is

$$Stock_t = \beta_1 + \beta_2 RTI_t + U_t$$

We can rewrite this expression as

$$U_t = Stock_t - \beta_1 - \beta_2 RTI_t$$

where U_t is the error term. If both *Stock* and *RTI* are cointegrated, then U_t should be stationary in its level.

If both *Stock* and *RTI* are co-integrated, a long-run equilibrium relationship exists between them, and in the short run there may be disequilibrium. Therefore one can treat the error term in the co-integration equation as the "equilibrium error." We can use this error term to tie the short-run behavior of *Stock* to its long-run value. Therefore this process is called "error correction."

If *Stock* and *RTI* are co-integrated, then causality must exist in at least one direction. To test the causality, the following VECM is estimated using OLS in each equation:

$$\Delta STOCK_t = \alpha_1 + \sum_{i=1}^{k} \beta_i \, \Delta RTI_{t-i} + \sum_{j=1}^{k} \lambda_j \, \Delta Stock_{t-j} + \omega_1 U_{t-1} + \varepsilon_{1t} \ldots \tag{6.3}$$

$$\Delta RTI_t = \alpha_2 + \sum_{i=1}^{k} \gamma_i \, \Delta RTI_{t-i} + \sum_{j=1}^{k} \pi_j \, \Delta Stock_{t-j} + \omega_2 Z_{t-1} + \varepsilon_{2t} \ldots \tag{6.4}$$

where Δ as usual denotes the first difference, ε_t is a random error term, $U_{t\text{-}1} = (Stock_{t\text{-}1} - \beta_1 - \beta_2\, RTI_{t\text{-}1})$, and $Z_{t\text{-}1} = (RTI_{t\text{-}1} - \beta_1 - \beta_2\, Stock_{t\text{-}1})$; that is, the one-period lagged value of the error from the co-integrating regression. The coefficients on the error correction component, ω_1 and ω_2, measure the rate at which disequilibria are corrected. If the estimated coefficients on lagged *RTI* in equation (3) are statistically different from zero as a group (i.e., $\sum \beta_i \neq 0$) and the set of estimated coefficients on lagged *Stock* in equation (4) is not statistically different from zero (i.e., $\sum \pi_j = 0$), then *RTI* leads *Stock*.

6.4.3 Back-Testing Results

Excluding private companies, we are generating RTI for a total 2761 publicly traded companies as of September 2014. Across the entire portfolio, we have so far found that RTI is a leading indicator of price movements in roughly about 15–20% of the companies. In the remaining companies, we have found no statistically significant relationships between RTI and security prices.

We present the results from a sample portfolio of oil and gas companies in our corpus. Of 137 companies in the portfolio, 31 (22.7%) showed a lead as of the end of August 2014 (Figure 6.3).

Every week, a sample of these content items is manually curated and compared against the system's automated results. We accumulate each week's

RTI Leads Stock	Oil & Gas Exploration & Production	Oil & Gas Refining & Marketing	Oil Related Services & Equipment	Total
1 Day Lead	16	5	2	23
More than 1 Day Lead	4	0	4	8
Total Companies with Lead	**20**	**5**	**6**	**31**
Companies with No Relationship				**106**
Companies with Lag				**0**
Total O&G Coverage				**137**

Figure 6.3 Sample oil and gas portfolio

sample, and compare the weekly tests against a growing sample of manually curated items. The system's accuracy is close to 85% in classification and 80% in impact assessment. The assessment is across sectors and spans a large number of factors.

Our analysts attempt to identify if any companies are dropping off the lead list or if there are new factors that need to be included among the operating models for the sector or any company. Any changes in the operating models, ***O***, are calibrated against a manually curated Golden Content Set using bootstrapping (Efron, 1993). The new operating model for the company, ***O***', is promoted to production use only if it shows improvement.

To give a brief idea of this assessment, in Figure 6.4 we present the RTI signals on a randomly selected company from the 31 companies with a lead in the sample oil and gas portfolio – Norfork Southern. According to our back-testing, Norfolk Southern's RTI signal shows a moderate lead over stock prices.

We also track analyst rating changes (ARC). They are shown as X's on the graph. Interestingly, in the period November 13 to February 14, most analyst ratings were "neutral" and in fact at least one firm downgraded NSC.N to "hold." Beginning on Feb '14, analyst ratings began to project a more positive outlook for NSC.N. RTI signals, however, stayed positive throughout the period. We additionally ran a different type of back-test to see if trading based on signals generated significant alpha after adjusting for transaction costs.

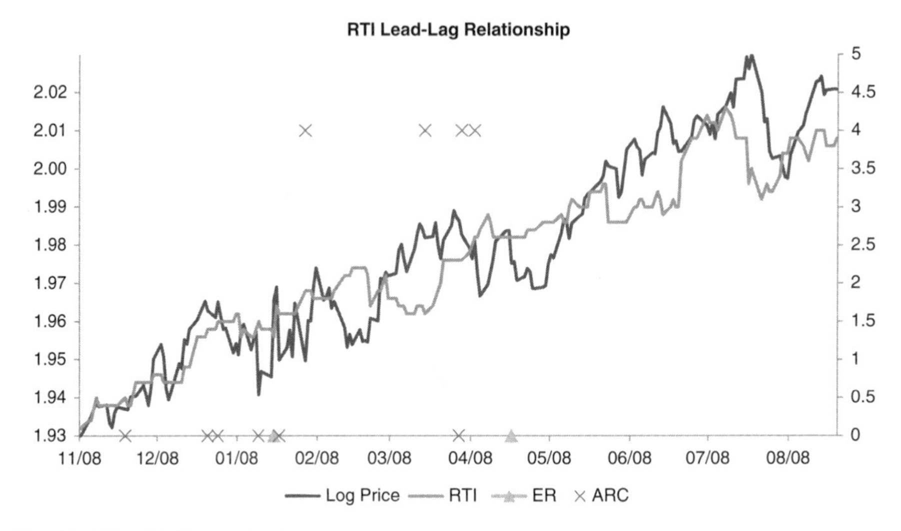

Figure 6.4 Impact analysis for Norfolk Southern

Trading Rules

1. We start with a phantom quantity, say 1000 shares that are bought in the beginning of the trading period.
2. We buy/sell shares depending on the change in cumulative score in the RTI.
3. Quantity in hand is changed on each buy/sell transaction.
4. If quantities in hand is negative (i.e., the investor does not have stock in hand), no transaction will take place (no margin trading).

Calculation of Gains/Losses

1. We calculate the net gain/loss from trading by following a simple trading strategy of buying the stock at the beginning of the trading period and selling the stock at the end of the trading period.
2. We calculate the net gain/loss from trading by following the RTI trading strategy where we buy/sell shares depending upon the change in cumulative score in the RTI.
3. The difference of 1 and 2 above gives us the trading gain/loss through RTI.

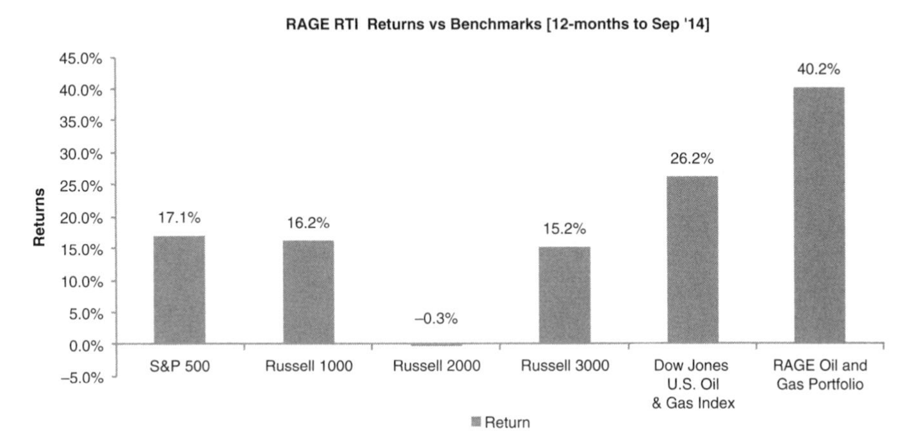

Figure 6.5 RTI trading returns versus benchmarks

Figure 6.5 presents the results of this trading strategy for a portfolio of 31 stocks in the oil and gas sector.

As the figure shows, the RTI-based trading clearly outperformed several indexes. When we calculated buy and hold returns on the same portfolio of 31 stocks, we also found that the returns for a buy and hold strategy were lower for the same portfolio over the same period.

The results in Figure 6.5 clearly tell us that there is likely value in systematically aggregating and interpreting information that can impact the intrinsic value of a firm. While this sample is small, a much larger sample showed similar results, with RTI returns higher than benchmarks.

6.5 SUMMARY

In this chapter, we have presented an intelligent system that leverages machine learning using computational linguistics and that generates alpha signals for financial markets. The system aggregates and interprets news and other content from around the world and, relying on an operating model for each company, determines the potential relevance and impact of the news and other content on each company's intrinsic value. Such relevance and impact are determined in a contextually relevant and causal manner and the platform's natural language understanding ability using each company's operating model, and so the system can process information at different levels of separation from the firm in terms of what it deems is in direct reference to the firm. By the type of system described in this chapter, active managers can move to an E4.0 state.

APPENDIX: SNAPSHOT OF THE OPERATING MODEL AT A SECTOR LEVEL FOR THE OIL AND GAS INDUSTRY

Integrated oil and gas			
	Industry drivers		
		Regional/global demand and supply	
			Demand and supply outlook
			Change in regional or global production
			Oil and gas industry
			Oil and gas services outlook
			Supply disruptions
			Crude oil prices
			Natural gas prices
			Product prices
		Supplementary industry outlook	
			Wind power outlook
			Biofuels outlook
			Petrochemical industry
			Petrochemical products
			Renewable energy industry
			Renewable energy demand
			Renewable energy by products
		Environmental regulations	

Note: Only three levels are displayed for brevity.

REFERENCES

Bhatia, S. (2007). Do the S&P CNX Nifty Index and Nifty Futures really lead/lag? Error correction model: A co-integration approach. NSE Working Paper, 1–31.

Brav, A., and Heaton, J. B. (2002). Competing theories of financial anomalies. *Review of Financial Studies* 15: 575–606.

Brooks, C., Rew, A. G., and Ritson, S. (2001). A trading strategy based on the lead-lag relationship between the spot index and futures contract for the FTSE 100. *International Journal of Forecasting* 17: 31–44.

Chan, W. S. (2003). Stock price reaction to news and no-news: Drift and reversal after headlines. *Journal of Financial Economics* 70: 223–260.

Easley, D., O'Hara, M., and Paperman, J. B. (1998). Financial analysts and information-based trade. *Journal of Financial Markets* 1: 175–201.

Efron, B., and Tibshirani, R. (1993). *An Introduction to the Bootstrap.* Boca Raton, FL: CRC Press.

Fama, E. F. (1998). Market efficiency, long-term returns, and behavioral finance. *Journal of Financial Economics* 49: 283–306.

Hartford Funds (2015). The cyclical nature of active & passive investing. White Paper, Hartford, CT.

Kim, O., and Verrecchia, R. (1994). Market liquidity and volume around earnings announcements. *Journal of Accounting and Economics* 17: 41–67.

Kim, O., and Verrecchia, R. (1997). Pre-announcement and event-period private information. *Journal of Accounting and Economics* 24: 395–419.

Marc, G., Timothy, K., and Wessels, D. (2005). Do fundamentals or emotions – drive the stock market? *The McKinsey Quarterly*, March.

RAGE Frameworks (2013). *A large scale system for contextually relevant analysis of unstructured content.* Rage White Paper. Dedham, MA.

Vega, C. (2006). Stock price reaction to public and private information. *Journal of Financial Economics* 82: 103–133.

CHAPTER 7

WILL FINANCIAL AUDITORS BECOME EXTINCT?

7.1 INTRODUCTION

External audits are a critical part of the financial framework underlying businesses worldwide. The integrity and thoroughness of such audits have been crucial in promoting investor confidence and in allowing commerce to flourish.

With the pervasive adoption of technology by enterprises, there has been significant interest in leveraging technology in order to make the audit process more efficient and to improve the quality of the audits. In fact, given the high degree of reliance on technology by corporations in general for running their business, it is imperative for audit firms to rely on automation to audit sophisticated technology systems adopted by their clients. Accounting standards now encourage audit firms to adopt IT and use IT specialists when necessary (AICPA 2002b, 2005, 2006b, PCAOB 2004b).

In terms of academic research on audit automation, even in the early days of computing, the potential impact of automation on accounting and auditing was recognized (Keenoy, 1958). Since then, there has been considerable academic research and focus on the subject, from both internal and external audit perspectives. Vasarhelyi (1984) pioneered research on the potential impact of audit automation on audit processes.

The Intelligent Enterprise in the Era of Big Data, First Edition. Venkat Srinivasan.

The focus of most academic research, and also commercial software, has been on point solutions and fairly rudimentary use of technology such as email Microsoft Office (Janvrin et al., 2008), and on specific aspects of the audit function, such as audit checklists by account and scheduling. The primary purpose of commercial software has been to assist in audit management, and not in automating the audit itself. The actual audit has remained a largely manual endeavor. However, even adoption of such basic productivity and audit management technology has had significant impact on the public accounting firm's productivity (Banker et al., 2002).

With the advances in business process automation and big data technology, there is a very real opportunity to intelligently automate the end to end audit lifecycle. The key to realizing the potentially enormous impact lies in the flexibility and adaptability of the solution.

In today's digital world, it is obvious that the effectiveness of any audit or risk assessment critically depends on the ability to process the enormous amount of information that is constantly arriving digitally. Auditors need to be well informed about their clients not just at the time of audit but on a continuous basis. We believe that the rate of arrival of new information far outstrips the capacity of audit firms and their staff to process such information.

This chapter outlines an adaptive, intelligent audit machine for the end to end, continuous external audit lifecycle. The solution is unique in its extensive flexibility and ability to continuously learn, automate audit examination, and adapt to changes in the audit processes, standards, and knowledge rapidly. The solution derives its learning ability, flexibility, and adaptability from the extensive highly model-driven architecture of the RAGE AI™ platform. The solution also unobtrusively incorporates two types of machine intelligence in audit processes. It incorporates auditable discovery of rules, rule-based reasoning, and expert knowledge for automated analysis of structured and deterministic data. For unstructured data, it incorporates the unique capabilities of RAGE AI™ for natural language understanding and automated machine learning in a highly contextual, auditable fashion that we believe is an absolute must in most decision-making contexts. The audit team has to be able to understand the precise reasons for any automated interpretation.

The solution is extensible. For example, while the solution does not explicitly address fraud detection, since fraud detection is not part of the external audit lifecycle, fraud detection algorithms or approaches can be easily integrated into the solution framework in a non-intrusive manner. The solution and approach can also be extended to many other types of audits, such as regulatory audits and internal audits.

7.2 THE EXTERNAL FINANCIAL AUDIT

While there are various industry and company specific variations of the audit process, the high level steps remain essentially the same. Figure 7.1 outlines a high-level fiduciary financial audit process. The process starts with the step of the audit firm accepting the client and ends with the issue of an independent opinion as to truth and fairness of the financial statements, by the audit firm.

7.2.1 Client Engagement

Prior to accepting a client, the audit firm is expected to ascertain that the audit firm is sufficiently independent of the client, its directors, and senior executives. This is to assure the integrity of the audit. Once independence is verified, the audit firm also ascertains whether the client is really ready for an external audit. In some cases, especially the bigger audit firms also perform background checks of key members of management.

7.2.2 Audit Planning

Once the engagement letter is executed, the audit firm needs to formally plan the audit. Audit planning for public companies is governed by PCAOB Auditing Standard No. 9. For non-public entities, auditors in the United States

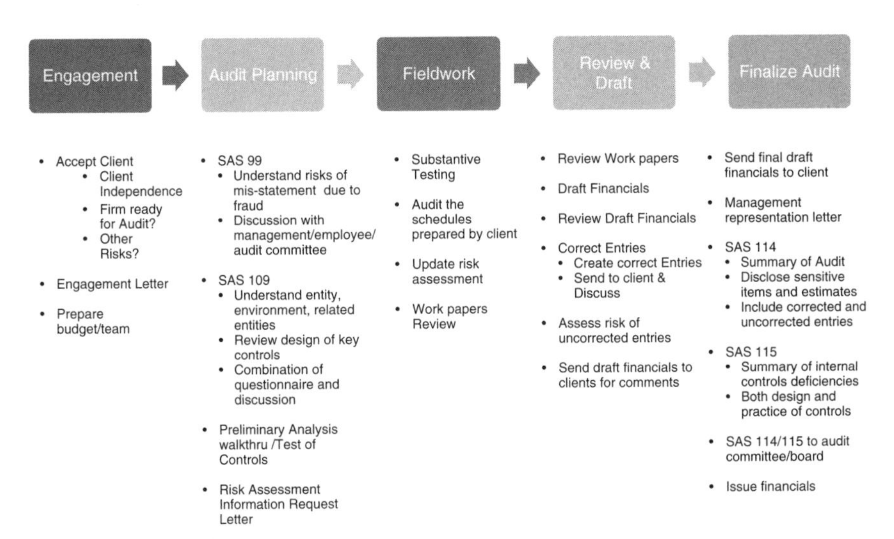

Figure 7.1 A generic external audit process*

*SAS standards are by AICPA for non-issuers; issuers are governed by equivalent standards from the PCAOB. Not listed here for brevity.

are expected to follow the guidelines in American Institute of Certified Public Accountants Statement of Auditing Standards No. 109 and 99. A lot of emphasis is placed by the regulatory bodies and the profession on audit planning.

Audit planning revolves around developing an in-depth understanding of the risks of material misstatements in the client's financial statements and to ascertain materiality limits. To develop such an in-depth understanding, the auditors have to become intimately familiar with the entity, its business, the industry it operates in, related entities, its past financial performance, and so on. This is typically done through a combination of visiting the client premises, by discussions with key management and personnel of the client in person and through questionnaires, assessment of the design of internal controls, and reviewing the company's legal and financial documents, including its charter.

Understanding the key financial performance indicators relative to the industry can allow the auditor to assess the reasonableness of client financial statements and to identify exceptional financial data. The auditors' assessment of risks of materiality and misstatements directly determines the nature, timing, magnitude, and emphasis in audit procedures. For example, the auditor may decide on unequal emphasis between internal control tests (walk through tests of internal control in actual transactions) and substantive tests (tests designed to verify and substantiate the completeness and accuracy of transactions and/or account balances).

7.2.3 Fieldwork

Fieldwork refers to the execution of audit procedures designed in the audit planning step. In the case of internal control testing, tests must be conducted to ensure that the controls have been and are actually working. An example internal control procedure is to have two signatories sign off when a wire transfer above a threshold is effected. The audit test of this control will be to select a random set of wire transactions above the threshold and verify physically that there were two authorized signatories on all of them.

Often substantive tests form the bulk of the fieldwork by auditors. Fundamentally, in these procedures, accounting data and their interrelationships are examined to determine the "reasonableness" of account balances. Each financial statement item amount or balance embodies a number of assertions. If the assertions are erroneous, then the financial statement amount contains misstatements.

Assertions are generally categorized into the following two categories. Assertions relating to transaction classes:

Attribute	Definition
Occurrence	Recorded transactions actually pertain to the entity
Completeness	No transactions have been left unrecorded
Accuracy	Transactions have been recorded properly and accurately
Cut-off	Transactions have been recorded in the correct accounting period
Classification	Transactions have been classified under proper accounts

Assertions about account balances as of the end of the audit period:

Attribute	Definition
Existence	Assets, liabilities and equity exist
Rights & Obligations	The entity's rights to the balances is not in question
Completeness	No transactions have been left unrecorded
Valuation	Assets, liabilities and equity have been properly valued

The audit objective is to ascertain the validity of each assertion relating to a class of transactions and account balances.

Substantive testing can be accomplished with a combination of two approaches:

1. *Substantive analytical procedures* In this approach, the auditor examines the interrelationships between account balances to ascertain their reasonableness.
2. *Tests of details* In this approach, the auditor examines the transactions that result in the account balance or examines the components of the closing account balance directly.

Substantive testing enables the auditor to verify the validity, completeness, and accuracy of the assertions implied by the financial statements.

7.2.4 Review and Draft

After substantive testing is completed, the auditor needs to appraise the findings:

1. Identify if there are contingent liabilities that need to be recorded.
2. Review subsequent events before and after the audit report is signed.
3. Re-assessment of financial statement validity based on the going concern assumption.
4. Assess the risks of uncorrected entries.
5. Review and evaluate audit evidence collected till now and working papers.

At this stage, the auditor prepares draft financial statements and sends them to the client for their comments.

7.3 AN INTELLIGENT AUDIT MACHINE

In this section, we describe an intelligent audit machine (IAM) for an external audit. The objective of IAM is to improve the quality and consistency across audits, usher in continuous audit, achieve a significant reduction in the overall effort associated with external audits for both the audit firm and client, and enable the audit firm to institutionalize audit standards.

The overall conceptual solution is presented in Figures 7.2, 7.3, and 7.4. To be more precise, Figure 7.2 presents the functional architecture for the solution, Figure 7.3 the IAM architecture using RAGE AI™, and Figure 7.4 the IAM architecture showing different pieces of the IAM. As can be seen in Figure 7.2, we envision a flexible and modular solution for each phase of the audit process. The modules correspond to the high-level stages in Figure 7.1.

We envision the IAM solution being used by audit teams, audit clients, and third parties who have a role in the audit process, for example, debtors confirming accounts receivables. We envision a self-serve audit portal that the client will use, an application for the audit teams and the audit firm, and finally, a portal for interaction with third parties, for example, banks, creditors, and debtors. The solution architecture is highly flexible because it is meta model driven. Using RAGE AI™, all the functionality is reduced to a set

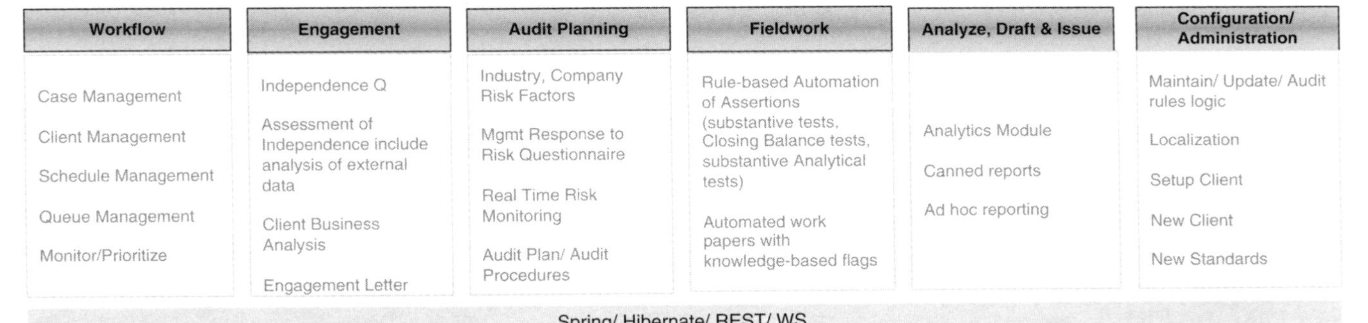

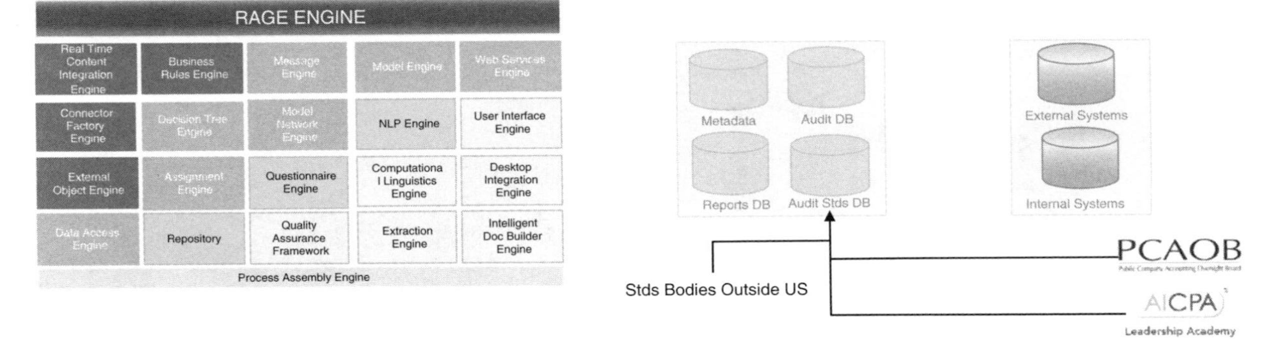

Figure 7.2 Intelligent audit machine – Functional architecture

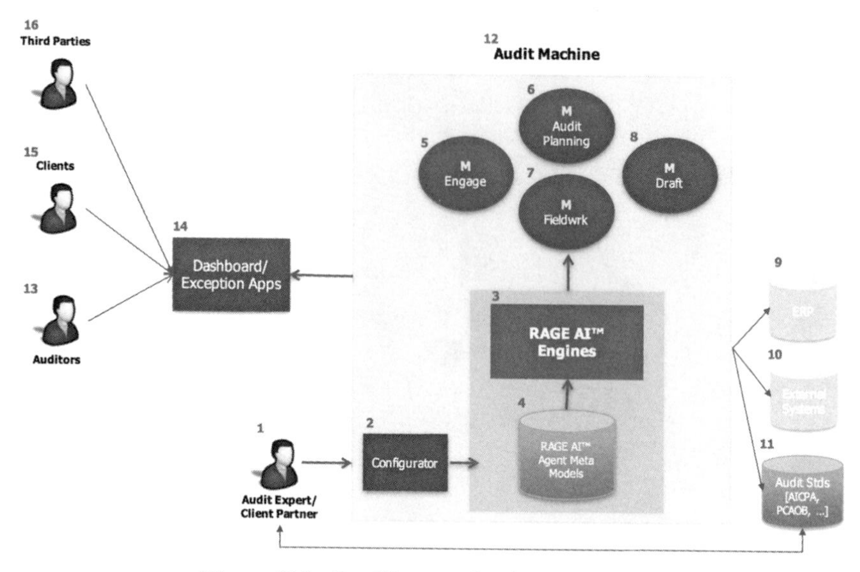

Figure 7.3 Intelligent, adaptive automation

of business process models in a no-code environment. These models can be modified with ease without any programming.

With the extensive set of knowledge-based components in RAGE, the IAM can gain knowledge on a continuous basis from both structured and

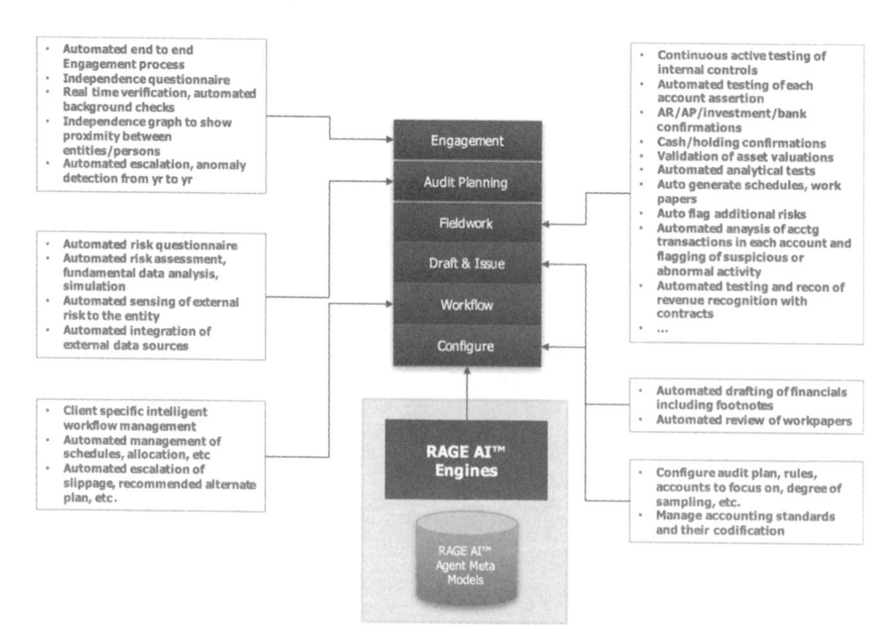

Figure 7.4 Intelligent, adaptive automation

unstructured data and with or without expert assistance. As audit teams and users identify new knowledge or process improvements, they can be implemented almost immediately. Such flexibility in institutionalizing knowledge gives the IAM the ability to adapt to changes in the client landscape, and to changes in standards and regulations.

Figure 7.3 outlines the IAM architecture and shows the set of automated agents that can be configured and deployed with the RAGE AI™ platform and components. These are illustrative. Additional automation agents can be configured quite easily with the platform. Many, if not all, audit processes can be completely automated leaving the audit teams to focus on understanding the risks related to the fair and reasonable statement of financial standing of the client and reaching an informed opinion.

7.3.1 Client Engagement

We envision the end to end engagement acceptance process to be substantially automated and potentially enhanced in the IAM solution. The questionnaire component can be used to set up relevant questions and auto-analysis of responses. The IAM solution can also enable the setup of client specific questions. IAM can auto-analyze the responses.

The IAM comprises a comprehensive set of secondary machines that are synchronized to conduct the audit using the audit design and plan provided by the audit expert such as the client partner. Figure 7.3 illustrates the functions of four sets of secondary intelligent machines in the IAM:

- Sub-machine 5 handles the engagement process.
- Sub-machine 6 manages and executes the audit planning process.
- Sub-machine 7 does all the fieldwork. This sub-machine will have many autonomous parts or agents that conduct the fieldwork tasks. There may be agents at the item level, such as an AR confirmations agent that manages and obtains all receivable confirmations, an inventory agent that electronically extracts inventory information, schedules physical verification if necessary, and does the reconciliation tasks.
- Sub-machine 8 generates the draft audit report including financials and footnotes.

Figure 7.3 also illustrates how the IAM is integrated with ERP or accounting systems where clients consent to such integration. Alternatively, of course, G/L and sub-ledgers can be digitally fed to the IAM. Audit standards and procedures can also be provided to the IAM. In a semi-supervised implementation of the IAM, such standards can be codified into an actionable form so that the IAM can test for compliance to standards.

Figure 7.4 illustrates in more detail the types of agents and automation that can be achieved within each area.

The solution can also automatically integrate with external systems that perform background checks. We also expect to add an independence network to examine if a change in one executive's or client situation would be consequential for other clients.

7.3.2 Audit Planning

Audit planning tasks consist of workflow and risk assessment. Workflow tasks can be supported by functionality related to scheduling, calendaring, budgeting, and resource allocation. There are several commercial products that provide this functionality.

However, risk assessment is an area of audit planning that the IAM can enhance significantly. In this regard we differentiate risk assessment from detection of fraud. The objective of the external audit does not, in the normal course, include the detection of fraud. The IAM may discover fraud but that is typically an unintended result. An automated journal entry agent would examine all journal entries and flag unusual or suspicious entries.

In the IAM, the audit team can implement analytical rules and logic using the extensive set of knowledge-based components in the RAGE platform, mainly rules, decision tree, model, and model network. Such logic can be at the account item level or at the client or industry level. Besides the ability to define rules and logic, an extensive set of fundamental metrics and risk assessment parameters can be automatically executed and assessed (Figure 7.5).

In using RAGE AI^{TM}'s natural language based extraction and normalization components, the IAM can extract and normalize fundamental data directly from source documents, such as corporate filings and annual reports.

Figure 7.5 Fundamental analysis

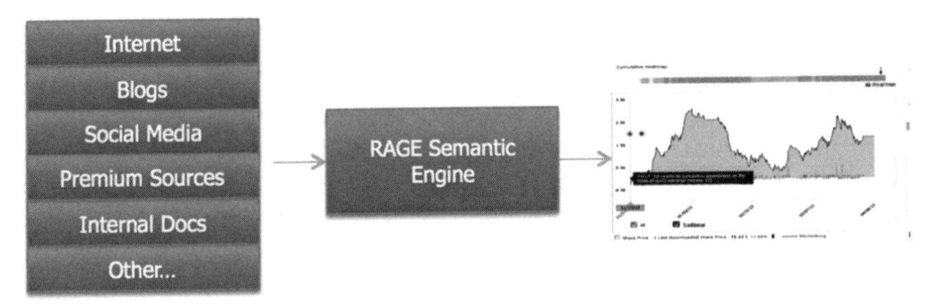

Figure 7.6 Continuous risk signals

For private sector companies, such data can be accessed from all the ledgers and sub-ledgers across divisions, subsidiaries, and normalized/consolidated as needed.

IAM can also assess entity risk continuously by integrating and analyzing external information such as news and social media posts. Figure 7.6 illustrates such analysis of unstructured information from around the world and the interpretation of unstructured information from a solvency point of view. This IAM automated sub-machine can generate a continuous signal so that risks can be incorporated at any time in the audit cycle.

IAM can be integrated with credit information sources in all countries. These sources transmit alerts on the businesses they monitor on a daily basis.

IAM can dynamically produce audit checklists by account and also generate information requests for the client. Depending on the client, the request can be either through the audit portal or through a printed document.

7.3.3 Fieldwork

Fieldwork consumes a huge amount of time in most audits. The audit team spends considerable effort in substantives tests of assertions. Fieldwork is also amenable to significant automation. As illustrated in Figures 7.3 and 7.4, the IAM can include as many secondary machines as necessary for specific items, such as AR, AP, inventory, and intangibles. Each of the sub-machines can handle the details of a specific account and feed the results to other sub-machines as needed.

7.3.4 Existence Tests

In financial accounts, the existence of assets and liabilities often involve third parties, such as debtors, creditors, banks, and custodians, and involve a confirmation of existence by the third party. This is referred to as the "confirmations" process. Most existence assertions can be automated through the IAM.

Typically, auditors verify a sample of the transactions in an account that leads to the closing balance of assets and liabilities. From intelligently selecting the sample to generating the confirmation and processing the response from the third party, the IAM can automate the entire process. In fact, the IAM can automatically process a lot more confirmations and not restrict itself to a sample as is done usually in the manual process. Through its patented extraction component, the process of matching confirmed amounts and other attributes with the recorded amounts and attributes can be automated.

7.3.5 Rights and Obligations

Auditors also review contractual documents to understand the rights and obligations of the client as a contracting party to any contracts. An important example is a software contract in the case of a software firm. The audit team will be interested in understanding terms of the contract such as software licensing fees, implementation or setup fees, liabilities as indicated by acceptance language, and any extent of warranties. All this is needed to determine the appropriateness of revenue recognition by the client.

The IAM solution is unique in its ability to automatically review contracts and reconcile the terms and conditions with other derived data and documents such as invoices and bills. RAGE AI™'s natural language understanding capabilities enable such automated analysis of contractual documents. Its capabilities go far beyond mere identification and extraction of key terms and extend to a qualitative assessment of different clauses or semantic matching of attributes with external data.

7.3.6 Substantive Analytical Procedures

The IAM can further provide automated assistance in analytical procedures where audit teams examine ratios relative to past trends, other firms in the industry and/or industry as a whole. Auditors can be alerted for unusual values.

7.3.7 Closing Balance Tests

The IAM can automatically create schedules and work papers at an account level identifying significant and unusual transactions in the process.

Figure 7.7 Audit standards actionable intelligence

7.3.8 Analyze and Issue Financials

We envision the IAM automatically generating a draft of the financials and audit report with all relevant footnotes. The IAM could also generate the management letter for internal control deficiencies. The audit team would then edit these documents on the system and issue them electronically.

7.3.9 Audit Standards

In today's largely manual world, audit standards are typically known to the auditors. In IAM, the audit firm can codify standards and make them more actionable (Figure 7.7). For example, often standards are expressed in a series of questions the answers that help determine the treatment of a specific item. Such standards can be codified into IAM whereby the IAM would not only identify/obtain the answers to such questions, but potentially arrive at a conclusion that is completely traceable. Auditors could then trace the reasoning of the IAM to their satisfaction. An example is the recent standard issued by the AICPA for revenue recognition in the case of Software as a Service (SaaS) contracts.

Such a process of codifying the firm's implementation of standards would lift the burden on auditors' having to scrutinize new and updated standards in order to operationalize them. Once the standards are codified using a metadata structure in RAGE AI™, they are effectively institutionalized and can influence every audit that is affected by the new standards.

Such a process is also likely to be viewed very favorably by the regulatory bodies.

7.3.10 Workflow/Configuration

IAM can already intelligently control and manage workflow to optimize resource allocation on all sides through various stages of the audit process. This functionality can address scheduling, calendaring, and resourcing type functions.

IAM is entirely configurable. From adding and setting up clients to defining interpretation rules for different sectors, industries, and account types to rules for text generation, the audit firm can create highly reusable metadata for automating client audits in different industries. New standards can be codified and mapped to the respective account set.

7.4 SUMMARY

There is consensus in academe and among practicing auditors that external and internal audits can benefit significantly with knowledge-based technologies. In this chapter, we have delineated our conception of a comprehensive intelligent machine for external financial audits. The intelligent audit machine that we have designed will dramatically change the way external audits are conducted. The depth and breadth of a typical audit will expand improving the quality of audits. Consistency will also improve across the board. Audit firms and clients will see significant reduction in effort for an audit. Audit firms will move to the Enterprise 4.0 state.

As discussed, the IAM can be extended to many other audit processes. Besides use by internal auditors, regulators can configure IAMs for their purpose substantially easing the regulatory burden on the companies being regulated and, at the same time, improving the quality of their work.

REFERENCES

Abdolmohammadi, M. J. (1999). A comprehensive taxonomy of audit task structure, professional rank and decision aids for behavioral research. *Behavioral Research in Accounting* 11: 51–92.

Alles, M. A., and Datar, S. (2004). Cooking the books: A management-control perspective on financial accounting standard setting and the Section 404 requirements of the Sarbanes–Oxley Act. *International Journal of Disclosure and Governance* 1 (2): 119–137.

American Institute of Public Accountants (AICPA) (2002). Consideration of Fraud in Financial Statement Audit. Statement of Auditing Standards No. 99. New York: AICPA.

American Institute of Public Accountants (AICPA) (2006) Audit Documentation. Statement of Auditing Standards No. 103. New York: AICPA.

American Institute of Public Accountants (AICPA) (2006). Risk Assessment Standards. Statement of Auditing Standards 104-111. New York\: AICPA.

Bamber, E. M., and Ramsey, R. J. (2000). The effects of specialization in audit work-paper review on review of efficiency and reviewers' confidence. *Auditing: A Journal of Practice and Theory* 19 (Fall): 147–157.

Banker, R. D., Chang, H., and Kao, Y. (2002). Impact of information technology on public accounting firm productivity. *Journal of Information Systems* 16 (2): 209–222.

Bell, T. B., Bedard, J. C., Johnstone, K. M., and Smith, E. F. (2002). KriskSM: A computerized decision aid for client acceptance and continuous risk assessment. *Auditing: A Journal of Practice and Theory* 21 (September): 97–113.

Elliott, R. K., and Jacobson, P. D. (1987). Audit technology: A heritage and a promise. *Journal of Accountancy* (May): 198–217.

Fischer, M. J. (1996). "Realizing" the benefits of new technologies as a source of audit evidence: An interpretive field study. *Accounting, Organizations and Society* 21 (February–April): 219–242.

Hunton, J. E., Wright, A. M., and Wright, S. (2004). Are financial auditors over-confident in their ability to assess risks associated with enterprise resource planning systems? *Journal of Information Systems* 18(2): 7–28.

International Accounting Bulletin (2005). Interview – BDO Seidman: Grabbing new opportunities. *International Accounting Bulletin*, London, September 8: 5.

Janvrin, D., Bierstaker, J., and Lowe, D. J. (2008). An examination of audit information technology use and perceived importance. *Accounting Horizons* 22 (1): 1–21.

Keenoy, C. L. (1958). The impact of automation on the field of accounting. *Accounting Review* 33 (2): 230–236.

Knechel, W. R. (1988). The effectiveness of statistical analytical review as a substantive accounting procedure: A simulation analysis. *Accounting Review* 63 (January): 74–96.

Leech, S. A., and Dowling, C. (2006). An investigation of decision aids in audit firms: Current practice and opportunities for future research. Working paper. University of Melbourne.

Manson, S., McCartney, S., and Sherer, M. (1997). *Audit Automation: The Use of Information Technology in the Planning, Controlling and Recording of Audit Work*. Edinburgh: Institute of Chartered Accountants of Scotland.

Manson, S., McCartney, S., Sherer, M., and Wallace, W. A. (1998). Audit Automation in the US and UK: A comparative study. *International Journal of Auditing* 2: 233–246.

Messier, W. F. Jr., and Hansen, W. A. (1987). Expert systems in auditing: The state of the art. *Auditing: A Journal of Practice and Theory* 7 (Spring) 94–105.

O'Donnell, E., and Schultz, J. (2003). The influence of business-process-focused audit support software on analytical procedures judgements. *Auditing: A Journal of Practice and Theory* 22 (September): 265–279.

Public Company Accounting Oversight Board (PCAOB) (2007). Auditing Standard No. 5. An Audit of Internal Control Over Financial Reporting That is Integrated with an Audit of Financial Statements.

RAGE Frameworks (2013). Real time credit risk assessment and monitoring. RAGE White Paper, April.

RAGE Frameworks (2013). On information diffusion in financial markets – Evidence from news, blogs and social media. RAGE White Paper, May.

Elkhoury, Marwan (2008). *Credit Rating Agencies and Their Potential Impact on Developing Countries*. United Nations Conference on Trade and Development. Discussion Paper 186, January. New York: UNCTAD.

Sprinkle, G. B., and Tubbs, R. M. (1998). The effects of audit risk and information importance on auditor memory during working paper review. *Accounting Review* 73 (October): 475–502.

Vera-Munoz, S. C., Ho, J. L., and Chow, C. W. (2006). Enhancing knowledge sharing in public accounting firms. *Accounting Horizons* 20 (2): 133–155.

Winters, B. I. (2004). Choosing the right tools for internal control reporting. *Journal of Accountancy* (February): 34–41.

INDEX

Note: Page numbers followed by "f" refer to figures.

The Intelligent Enterprise in the Era of Big Data, First Edition. Venkat Srinivasan.